Grade 4

Discovery Education | Science Techbook

Unit 1

Car Crashes

To obtain permission(s) or for inquiries, submit a request to:

Discovery Education, Inc.
4350 Congress Street, Suite 700
Charlotte, NC 28209
800-323-9084
Education_Info@DiscoveryEd.com

ISBN 13: 978-1-68220-798-7

Printed in the United States of America.

6 7 8 9 10 CWM 26 25 24 23   B

Acknowledgments

Acknowledgment is given to photographers, artists, and agents for permission to feature their copyrighted material.

Cover and inside cover art: Volodymyr Baleha / Shutterstock.com

Table of Contents

Concept 1.3 Speed

Concept 1.4 Energy and Collisions

Unit Wrap-Up

Grade 4 Resources

Dear Parent/Guardian,

This year, your student will be using Science Techbook™, a comprehensive science program developed by the educators and designers at Discovery Education and written to the Next Generation Science Standards (NGSS). The NGSS expect students to act and think like scientists and engineers, to ask questions about the world around them, and to solve real-world problems through the application of critical thinking across the domains of science (Life Science, Earth and Space Science, Physical Science).

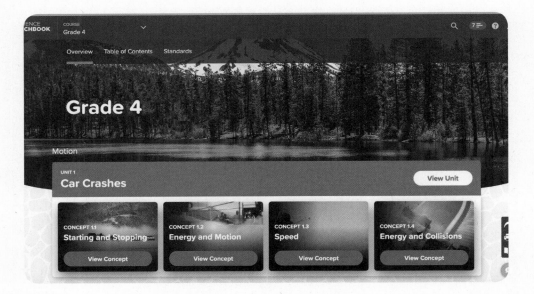

Science Techbook is an innovative program that helps your student master key scientific concepts. Students engage with interactive science materials to analyze and interpret data, think critically, solve problems, and make connections across science disciplines. Science Techbook includes dynamic content, videos, digital tools, Hands-On Activities and labs, and game-like activities that inspire and motivate scientific learning and curiosity.

You and your child can access the resource by signing in to www.discoveryeducation.com. You can view your child's progress in the course by selecting the Assignment button.

Discovery
EDUCATION

Science Techbook is divided into units, and each unit is divided into concepts. Each concept has three sections: Wonder, Learn, and Share.

Units and Concepts Students begin to consider the connections across fields of science to understand, analyze, and describe real-world phenomena.

Wonder Students activate their prior knowledge of a concept's essential ideas and begin making connections to a real-world phenomenon and the **Can You Explain?** question.

Learn Students dive deeper into how real-world science phenomenon works through critical reading of the Core Interactive Text. Students also build their learning through Hands-On Activities and interactives focused on the learning goals.

Share Students share their learning with their teacher and classmates using evidence they have gathered and analyzed during Learn. Students connect their learning with STEM careers and problem-solving skills.

Within this Student Edition, you'll find QR codes and quick codes that take you and your student to a corresponding section of Science Techbook online. To use the QR codes, you'll need to download a free QR reader. Readers are available for phones, tablets, laptops, desktops, and other devices. Most use the device's camera, but there are some that scan documents that are on your screen.

For resources in Science Techbook, you'll need to sign in with your student's username and password the first time you access a QR code. After that, you won't need to sign in again, unless you log out or remain inactive for too long.

We encourage you to support your student in using the print and online interactive materials in Science Techbook, on any device. Together, may you and your student enjoy a fantastic year of science!

Sincerely,

The Discovery Education Science Team

Discovery
EDUCATION

Unit 1
Car Crashes

The Science of Car Crashes

Lots of things happen in a car crash. There's a lot of noise. Things get broken and thrown about. People get hurt. This video shows one way in which car crashes are studied. We will learn about the things that can happen in a crash and what causes them as we study this unit.

Quick Code:
us4006s

Video

Science of Car Crashes

Think About It

Look at the photograph. **Think** about the following questions.

- What happens to energy when objects collide?

- Why do car crashes cause so much damage?

Car Crash

Solve Problems Like a Scientist

Unit Project: Self-Driving Cars

In this project, you will use what you know about energy, motion, and collisions to consider whether various effects of self-driving cars are beneficial or harmful. You will also examine the ethics and laws around new technologies like self-driving cars.

Quick Code:
us4007s

Car

| SEP | Developing and Using Models |
| SEP | Engaging in Argument from Evidence |

Discovery
EDUCATION

Ask Questions About the Problem

You are going to evaluate the benefits and possible dangers of self-driving cars. **Write** some questions you can ask to learn more about the problem. As you learn about energy, motion, and collisions in this unit, **write** down the answers to your questions.

Starting and Stopping

discoveryeducation.com

DISCOVERY
EDUCATION

Student Objectives

By the end of this lesson:

☐ I can explain and model what causes objects to change motion.

☐ I can analyze data to explain different causes of changes in an object's motion.

☐ I can cite evidence to show how speed is related to energy for an object.

☐ I can model the cause-and-effect relationship between the force acting on an object and the object's motion.

Key Vocabulary

☐ energy   ☐ gravity   ☐ rotate

☐ force   ☐ motion   ☐ work

☐ friction   ☐ resistance

Quick Code:
us4008s

Activity 1

Can You Explain?

How do forces act on a starting and stopping object?

_____

_____

_____

_____

_____

_____

_____

_____

_____

_____

_____

Quick Code:
us4010s

Activity 2

Ask Questions Like a Scientist

Takeoff and Landing

Watch the video of how planes take off and land. How does the plane stop?

Quick Code:
us4011s

Video

Let's Investigate Takeoff and Landing

SEP   Asking Questions and Defining Problems

What makes the aircraft move? How are they launched and stopped during landing? **Write** three questions you have, and **share** them with the class.

I wonder...

I wonder...

I wonder...

Activity 3

Observe Like a Scientist

Making Things Move

Watch the videos. **Look** for objects that change in motion.

Quick Code:
us4012s

How fast did it go?

Moving Things

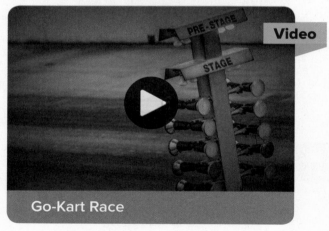

Go-Kart Race

Talk Together

Now, talk together about how the objects move in the videos.
How did forces cause the objects to move?

CCC   Cause and Effect

© Discovery Education | www.discoveryeducation.com ● Images: (a) Michael Boehler / EyeEm / Getty Images, (b) SchoolMedia, Inc., (c) Avigator Fortuner / Shutterstock.com, (d) Icon made by Freepik from www.flaticon.com

Activity 4
Evaluate Like a Scientist

What Do You Already Know About Starting and Stopping?

Always Changing

Look at the words that describe motion. **Circle** the word describing something that always changes when an object is in motion.

acceleration

direction

position

speed

CCC **Energy and Matter**

How Do Objects Move?

Pushing and pulling move objects. **Write** one sentence that describes pushing something. **Write** a second sentence that describes pulling something.

_____

_____

_____

_____

Balanced and Unbalanced

Look at the images showing a rope pulled in tug-of-war. **Circle** the image that shows the rope when it is not moving.

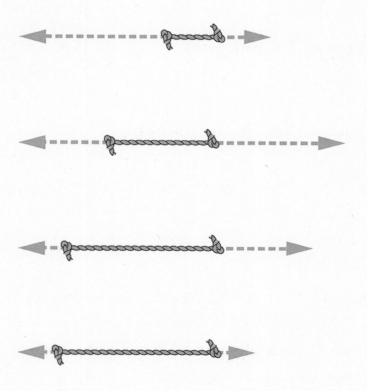

Activity 5

Analyze Like a Scientist

Objects in Motion

Quick Code:
us4014s

Read the text. Then, **write** your answers to the questions that follow.

Objects in Motion

An object is in **motion** if it is moving from one place to another. When you look at an object, you can describe its position compared to other things around it. For example, the starting position for your pencil might be the top right corner of your desk, next to an eraser. Then, it might roll down the desk until it hits your hand and stops. It now has a different position relative to the eraser. Since the pencil changed position, it was in motion relative to the eraser. For motion to start or stop, there must be a **force**, which is a push or a pull. **Gravity** pulled the pencil into motion, and your hand pushed on the pencil to stop it.

CCC **Energy and Matter**

Some motion is easy to see, and some is not. It is easy to see a person walk down the street, a leaf blow in the wind, or your pencil roll down the desk. However, you cannot see Earth move around the sun. You cannot see your desk **rotate** around the center of Earth. Motion is based on a person's perspective. For example, a satellite in space is able to observe Earth's rotation. But on Earth, because the other objects around us move with us as the planet rotates, it is not apparent that everything is moving together. You know an object is in motion if you can measure changes in its position, even if you cannot see those changes. An object's change in position is compared to something else, usually something that is not moving.

Think about tossing a ball in the air. Use what you learned from the reading to answer the questions. **Write** your answers in the space provided.

How would you know the ball is in motion?

_____

_____

_____

_____

_____

_____

What are two forces that can be exerted on the ball to make it move?

_____

_____

_____

_____

_____

_____

_____

Activity 6
Observe Like a Scientist

Motion

Quick Code:
us4015s

Watch the video about motion. **Look** for clues that show how an object is moving. Then, **write** your response to the statement.

Motion

Describe how you can tell that the bowling ball is in motion.

_____

_____

_____

_____

_____

_____

_____

What Makes Objects Move?

Activity 7
Observe Like a Scientist

Force

There are two types of forces: pushes and pulls. All around you, there are examples of these forces.

Watch the video on force.
Then, **answer** the questions.

Quick Code:
us4016s

Video

Force

What are some examples of exerting force with a push?

_____

_____

_____

_____

_____

_____

CCC **Cause and Effect**

© Discovery Education | www.discoveryeducation.com ● Images: (a) Michael Boehler / EyeEm / Getty Images, (b) Ihor Berkyta / Shutterstock.com

What are some examples of exerting force with a pull?

_____

_____

_____

_____

_____

Think about a force in your life. What would it be like if this force were no longer there?

_____

_____

_____

_____

_____

Activity 8
Observe Like a Scientist

Quick Code:
us4017s

Tug-of-War

Look at the image of a game of tug-of-war.

Tug-of-War

Draw what the tug-of-war would look like if both teams were equally matched.

CCC Cause and Effect

Activity 9
Analyze Like a Scientist

Stopping Motion

Quick Code:
us4018s

Before you **read** the text, **look** at these words and phrases. **Think** about what the text will be about based on this list. Then, **answer** the question that follows.

- slow down
- force
- space

- moving objects
- stop
- friction

Stopping Motion

Imagine you are standing on a spacecraft in deep space, and you throw the ball out into space. What do you think would happen?

There is no air to slow it down. There is nothing for it to hit. Will it stop, or will it go on forever? It may come as a shock to learn that the ball can travel on forever, unless a force slows it down—which is unlikely in space.

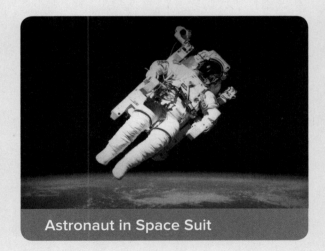

Astronaut in Space Suit

CCC   **Cause and Effect**

Stopping Motion *cont'd*

Moving objects only stop when a force of the same size is applied to them in the opposite direction from which they are moving. Sometimes it is easy to observe where the force that stops an object comes from. If a car crashes into a wall, it may stop. The wall applied a force to the car.

But why does that same car roll slowly to a stop if it runs out of gas on a level road? In this case, the car is being slowed down by a force called **friction**. You have probably heard of friction. Friction is a force that is exerted when objects rub against each other. Friction is a force that opposes motion. In the case of the car, this includes when its tires rub on the road and when air flows over the car and rubs against its surface.

Car Crash

When a car runs into a wall, what can you assume about the size of the force of the car compared to the size of the force of the wall?

_____

_____

_____

_____

_____

_____

Activity 10

Evaluate Like a Scientist

Quick Code:
us4019s

Mission to an Asteroid

How would a scientist explain the forces acting on a space probe? **Read** the text. **Circle** the correct word to complete each sentence.

Before launch, the Osiris-Rex space probe stays on Earth. It stays still because forces acting upon it are **balanced/pushing it down/ pushing it up/unbalanced**. A rocket can move the probe away from Earth. The rocket applies **balanced forces/unbalanced forces/ the force of friction/the force of gravity** to launch the probe into space. Once the probe is in space, there is no **balanced force/force of gravity/force of friction/unbalanced force**. The probe no longer needs the rocket to move. The probe will continue to move toward the asteroid because the probe has **no unbalanced forces/forces of gravity/forces of motion** to slow it down or change its path.

SEP **Constructing Explanations and Designing Solutions**

Activity 11

Investigate Like a Scientist

Quick Code:
us4020s

Hands-On Investigation: Rolling Cars

In this activity, you will explain how far cars travel when different push forces are used. **Predict** how far the toy car or truck will roll. **Complete** the activity. Then, **record** your answers to the questions about the activity.

Make a Prediction

Write your claim here.

_____

_____

_____

_____

_____

_____

_____

_____

SEP Constructing Explanations and Designing Solutions

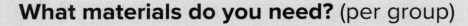

What materials do you need? (per group)

- Toy trucks
- Measuring tape
- Toy trucks, pull back
- Pull back car

What Will You Do?

1. Grab your toy cars and trucks.
2. Figure out a way you will measure the distance your toy cars will go.
3. Push a toy car hard.
4. Record the distance the toy car rolls.
5. Repeat steps 3 and 4 several times, and find the average.
6. Predict what will happen if you push your toy car very gently.
7. Record the distance the toy car rolls.
8. Repeat step 6 several times, and find the average.

Record your data in the table.

Trial	Type of Push	Distance
1	Hard	
2	Hard	
3	Hard	
4	Hard	
Average hard push distance		
5	Gentle	
6	Gentle	
7	Gentle	
8	Gentle	
Average gentle push distance		

Think About the Activity

Think about the data you collected. How does this data support or go against your hypothesis? **Describe** how you know. Then, **answer** the question.

My Claim _____

_____

_____

My Claim Is True Because

Could the distance each car traveled have changed if you had used a different car or truck?

_____

_____

_____

_____

_____

_____

_____

_____

_____

 Talk Together

What do you think caused the car to start and stop moving? What is your evidence? How does the car compare to the airplane you saw in Wonder?

Activity 12

Investigate Like a Scientist

Quick Code:
us4021s

Hands-On Investigation: Changing Motion

In this activity, you will explore how energy is related to force. **Predict** how speed and motion change when two balls collide. **Complete** the activity. Then, **record** your answers to the questions about the activity.

Make a Prediction

Write your claim here.

_____

_____

_____

_____

_____

_____

_____

_____

_____

_____

_____

_____

SEP **Constructing Explanations and Designing Solutions**

What materials do you need? (per group)

- Golf balls
- Ping pong balls
- Baseballs
- Whiffle balls
- Tennis balls
- Lacrosse balls

- Wooden board
- Several books
- Metric ruler
- Camera

What Will You Do?

1. Set up your ramp to move the ball.

2. Measure the mass or weight of the ball.

3. Place a stationary ball at the bottom of the ramp to explore what happens when the balls collide.

4. Push a ball down the ramp.

5. Describe the speed of movement as slow, medium, or fast.

6. Use the ruler to measure how far each ball moves after colliding.

7. Also measure the distance traveled from start to finish.

8. Change the height at which the ball rolls down the ramp.

9. Repeat steps 4–7.

10. Collect data and analyze results.

Think About the Activity

Think about the data you collected. How does the mass of a ball affect the amount of energy it has? Use what you observed to answer the question.

My Claim _____

_____

_____

My Claim Is True Because

What forces affected the speed and direction of the moving ball before it hit the other ball?

_____

_____

_____

_____

_____

Force and Energy

Quick Code:
us4022s

There is a relationship between force and energy. **Read** the text about force and energy. Then, **answer** the question.

Force and Energy

To make a vehicle move or stop requires a force—a pull or a push. To exert a force requires **energy**. Imagine you had to push a car along a flat road. Moving a car needs a lot of force. Soon you would be sweating hard as your body used up its energy reserves working to get the car moving.

Force and energy are different, but they are related to one another. Force is something that changes energy in such a way that it can do **work**. In the case of your pushing the car, the force your body exerts on the car is changing the energy in your body to energy in the moving car. When you move the car, you are doing work. To put it another way, a force transfers energy from one object to another. Work is the energy transferred by a force that is used to move the object.

SEP **Constructing Explanations and Designing Solutions**

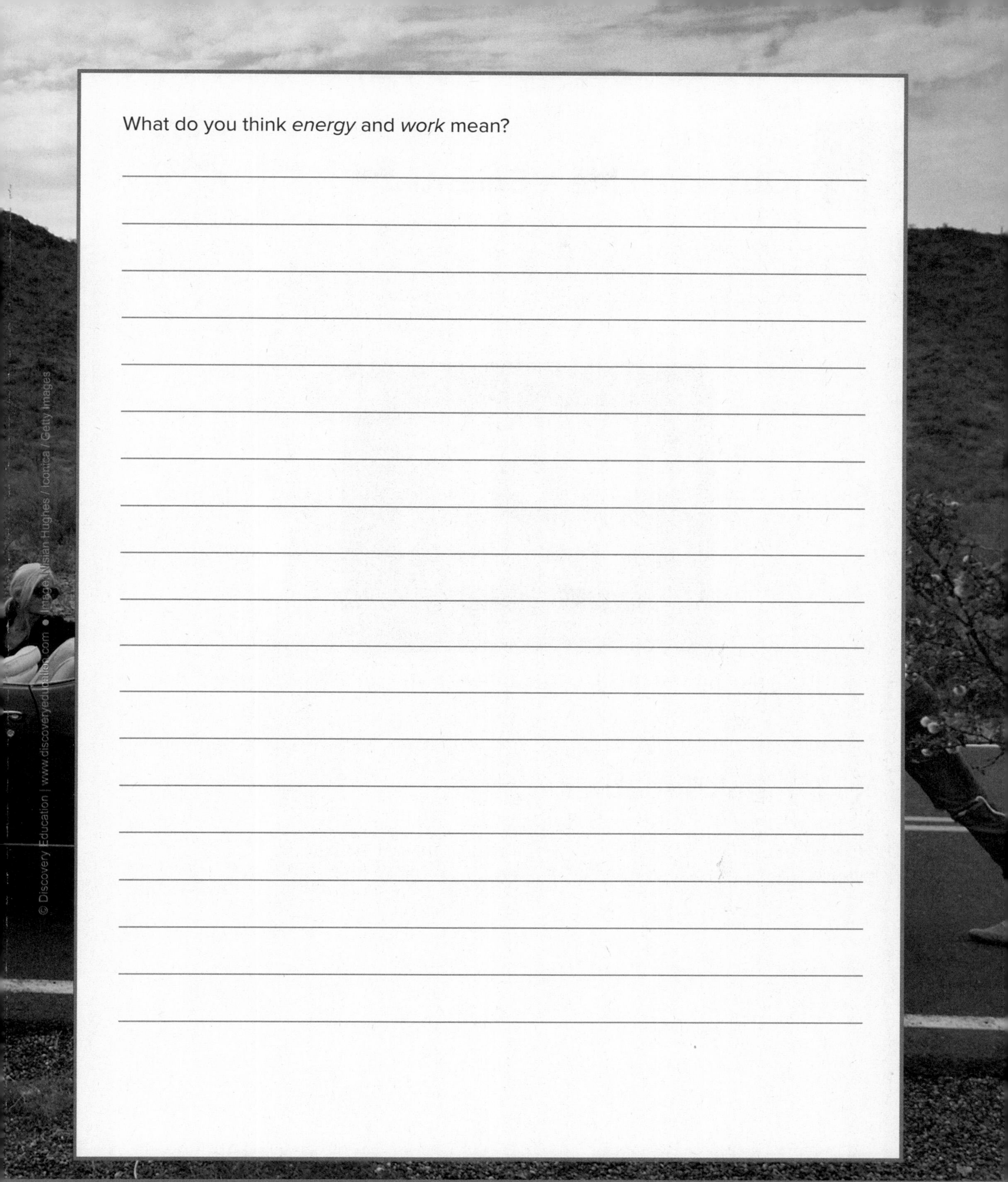

What do you think *energy* and *work* mean?

_____

_____

_____

_____

_____

_____

_____

_____

_____

_____

_____

_____

_____

_____

_____

_____

Activity 14

Observe Like a Scientist

Energy, Work, and Force

Watch the video to learn how energy, work, and force are related.

Energy, Work, and Force

Talk Together

Now, talk together about the nature of force, work, and energy.
What examples have you encountered during class?

Activity 15

Evaluate Like a Scientist

Quick Code:
us4024s

All Aboard!

Look at the picture. **Think** about these questions: How do you know the train was in motion? What force(s) do you think caused the train to start moving? What force(s) do you think caused the train to stop moving?

Commuter Train

Draw a picture of a train, and **use** an arrow to represent the force moving the train. **Label** your arrow with the name of the force.

CCC **Energy and Matter**

Activity 16

Record Evidence Like a Scientist

Quick Code:
us4025s

Takeoff and Landing

Now that you have learned about starting and stopping, watch the video on takeoff and landing again. You first saw this in Wonder.

Video

Let's Investigate Takeoff and Landing

Talk Together

How can you describe forces now?

How is your explanation different from before?

SEP **Constructing Explanations and Designing Solutions**

© Discovery Education | www.discoveryeducation.com ● Images: (a) Michael Boehler / EyeEm / Getty Images, (b) David M. Albrecht / Shutterstock.com, (c) Icon made by Freepik from www.flaticon.com

Look at the Can You Explain? question. You first read this question at the beginning of the lesson.

Can You Explain?

How do forces act on a starting and stopping object?

Now, you will use your new ideas about forces to answer a question.

1. **Choose** a question. You can use the Can You Explain? question or one of your own. You can also use one of the questions that you wrote at the beginning of the lesson.

My Question

2. Then, **use** the graphic organizers on the next pages to help you answer the question.

To plan your scientific explanation, first **write** your claim.

My claim:

_____

_____

_____

_____

_____

Next, **look** at your notes in the Bubble Map. **Identify** two pieces of evidence that support your claim:

Evidence 1

Evidence 2

Bubble Map

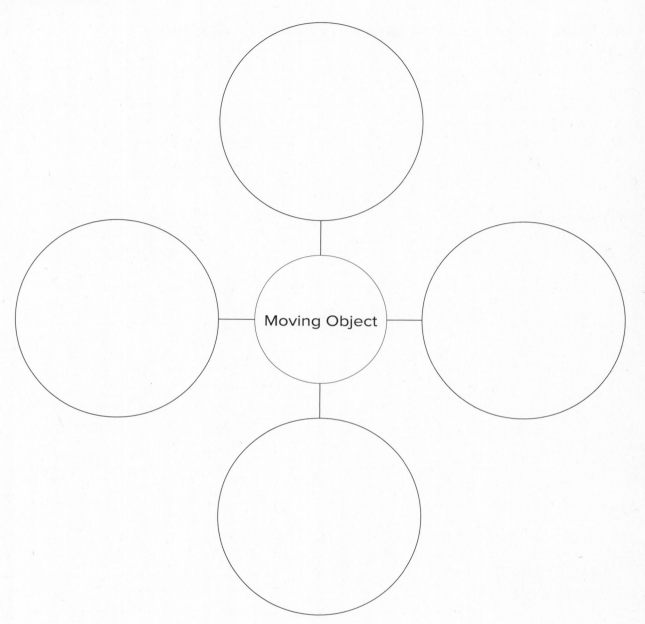

Moving Object

Now, **write** your scientific explanation.

The forces acting on a stationary airplane cause it to move because...

_____

_____

_____

_____

_____

_____

_____

_____

_____

_____

_____

_____

_____

_____

_____

_____

_____

STEM in Action

Quick Code:
us4026s

Activity 17

Analyze Like a Scientist

Launching an F-18

Read the text, and **watch** the video. Pay special attention to the training a U.S. Navy pilot must go through before flying a plane.

Launching an F-18

Have you seen video or photos of planes being launched from the deck of a ship? A team of people must do their job so that a plane can take off safely. Before the plane takes off, several people check it over to make sure all its parts are working properly. Each person has a routine to follow to keep planes safe.

The plane captain tells the pilot about any repairs made to the plane. Then, the pilot completes a check by walking around the plane. After the pilot is seated in the plane, other people on the ship's deck help the pilot make sure the plane works correctly. Finally, the plane moves to the location where it will launch from the ship.

Video

F-18 Pilot

To track the launch process, it is vital to know who does what. The crew knows a person's job from the color of the person's shirt. The crew also uses a variety of signals to communicate with the pilot.

U.S. Navy pilots go through a sequence of training before they fly:

- Officer Candidate School (OCS)

- Naval Aviation Schools Command

- Six weeks of primary flight training

- Intermediate flight training

- Assignment to a squadron

- Final training on the type of aircraft they will fly

Force on an Aircraft

How would a pilot explain the forces that act on an aircraft before it launches? **Read** the text. **Circle** the correct word or phrase to complete the sentence.

Due to **the force of friction/balanced forces/the magnetic force/unbalanced forces,** the plane remains at rest, sitting on the deck of the ship.

When the plane is waiting for takeoff, the forces acting on it are **balanced/ unbalanced/decreasing/increasing**.

When the plane takes off, the forces acting on it are **balanced/unbalanced/ decreasing/increasing**.

When the plane is in the air, the movement of the wind changes the forces that cause the plane to **change direction/move forward/slow down/stop**.

Becoming a Pilot

Think about the text you read and video you watched. What challenges do you think you would face when training to be a pilot? What would be the reward of this job? **Write** your answer to these questions.

_____

_____

_____

_____

_____

_____

_____

_____

_____

_____

_____

_____

_____

_____

Activity 18

Evaluate Like a Scientist

Review:
Starting and Stopping

Quick Code:
us4027s

Think about what you have read and seen in this lesson. **Write** down some core ideas you have learned. **Review** your notes with a partner. Your teacher may also have you take a practice test.

_____

_____

_____

_____

_____

_____

_____

_____

_____

_____

_____

_____

_____

_____

_____

SEP Obtaining, Evaluating, and Communicating Information

Talk Together

Think about the Science of Car Crashes video you saw in Get Started. Use your new ideas to discuss why forces cause things to move or stop.

Energy and Motion

Student Objectives

By the end of this lesson:

☐ I can investigate the forms of energy in a system or for an object.

☐ I can apply logical reasoning to predict the types of energy for an object.

☐ I can cite evidence to explain how energy is conserved.

Key Vocabulary

☐ chemical energy

☐ conservation of energy

☐ gravitational potential energy

☐ kinetic energy

☐ nuclear energy

☐ potential energy

☐ solar energy

☐ thermal energy

Quick Code: us4028s

Activity 1

Can You Explain?

How do moving objects get energy?

_____

_____

_____

_____

_____

_____

_____

_____

_____

_____

_____

_____

Quick Code:
us4030s

Activity 2

Ask Questions Like a Scientist

Quick Code:
us4031s

Roller Coasters

Watch the video about roller coasters. **Think** about what is needed to make them move.

Video

Let's Investigate Roller Coasters

What do you wonder about the energy used to get the roller coaster moving? What happened to that energy as it moved? **Write** three questions you have, and **share** them with the class.

I wonder...

I wonder...

I wonder...

Activity 3

Think Like a Scientist

Quick Code:
us4032s

Energy in the Classroom

Can you think of an example where energy is used? This word has different meanings.

In this investigation, you will explore the classroom in search of objects that use or contain different forms of energy.

What Will You Do?

Explore your classroom. Locate different objects that use or contain energy. Record what you find in the table. If you do not think an object uses energy, leave that cell blank. If you do not think an object contains energy, leave that cell blank. Remember that objects may use or contain several different kinds of energy.

Record your notes in the table.

Object	How Does It Use Energy?	How Does It Contain Energy?

Think About the Activity

Were you surprised at the number of objects you found that contained or used energy? Why?

_____

_____

_____

_____

_____

What were some forms of energy you observed?

_____

_____

_____

_____

_____

How did constructing a table help you to understand forms of energy?

_____

_____

_____

_____

_____

| SEP | Obtaining, Evaluating, and Communicating Information |

Activity 4

Evaluate Like a Scientist

What Do You Already Know About Energy and Motion?

Quick Code:
us4033s

Defining Energy

Write your own definition of *energy*. Include an example to support your answer.

_____

_____

_____

_____

_____

_____

_____

_____

_____

_____

_____

_____

_____

CCC **Energy and Matter**

Moving Energy

Think about boiling noodles on the stove. Next, **look** at the pictures.
Rank them from 1 (first) to 4 (last) to show how heat moves to cook the noodles.

Uncooked Noodles

_____

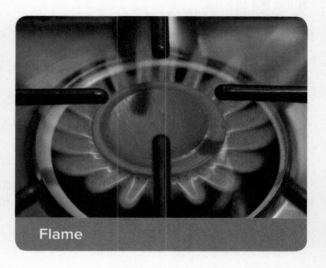

Flame

_____

Pan

_____

Water

_____

Activity 5

Analyze Like a Scientist

Quick Code:
us4034s

Kinetic and Potential Energy

Look at the image. **Think** about what would happen next. Then, **read** the text and **write** your answer to the question.

Potential and Kinetic Energy
Acrobat on the Tower Has Potential Energy

Potential Energy in Acrobats

CCC **Energy and Matter**

Kinetic and Potential Energy

Energy is the ability to do work. Energy makes things happen. If there were no energy on Earth, nothing would get done! There are two categories of energy: **kinetic energy** and **potential energy**. Kinetic energy is the energy of motion. The word *kinetic* means that something is moving. In other words, kinetic energy is the energy an object has because it is moving. Potential energy is stored energy or the energy of position. The word *potential* means that something is possible. In other words, potential energy means that an object is ready to do work or to be active.

What most likely happened next in the picture of the four acrobats?

_____

_____

_____

_____

_____

Activity 6
Observe Like a Scientist

Energy Basics

Quick Code:
us4035s

Watch the video. **Look** for forms of energy that are visible and invisible. **Share** with your class how work and energy are related to one another.

Video

Energy Basics

Talk Together

Now, talk together about the ways work and energy are related.

Activity 7

Analyze Like a Scientist

Conservation of Energy

Quick Code:
us4036s

Think about how many different forms of energy you can name.
Read the text, and **write** a question based on what you read. **Share**
your question with a partner to see if he or she can answer it.

Conservation of Energy

When energy changes forms, no energy
disappears and no new energy is made.
This means that energy is conserved. The
amount of energy stays the same. The
sun **converts** a lot of potential energy
into **solar energy**. The potential energy
is the **nuclear energy** and is released
by the particles that make up the sun as
solar energy.

Fire Burning

Plants use solar energy to grow. Plants grow into trees that may be cut down
to become firewood. When firewood is burned, **chemical energy** is changed
first into **thermal energy** and then into light and heat.

CCC **Energy and Matter**

Does the energy get destroyed? No, it does not. It just travels to the atmosphere and then out to space!

The chemical energy in fireworks is changed into thermal, kinetic, light, and sound energies when the fireworks are lit and explode. If you could measure the amounts of these energies, you would find that the amount of chemical energy is the same as the amount of thermal, kinetic, light, and sound energies combined. This is because energy is conserved. It is neither created nor destroyed but simply changes forms.

Write one question you have after reading the text.

_____

_____

_____

_____

_____

_____

Activity 8

Observe Like a Scientist

Law of Conservation of Energy

Quick Code:
us4037s

Watch the video about how energy is conserved. **Look** for different forms of energy. Then, **answer** the questions.

Law of Conservation of Energy

What does **conservation of energy** mean?

_____

_____

_____

Write two examples of one form of energy being transformed into another form.

_____

_____

_____

_____

CCC **Energy and Matter**

What Are the Different Forms of Potential and Kinetic Energy?

Activity 9
Analyze Like a Scientist

Forms of Potential and Kinetic Energy

Quick Code:
us4038s

Read the text. Then, **answer** the questions that follow.

Forms of Potential and Kinetic Energy

Potential energy is energy that is stored in an object. You could say that an object with potential energy is not doing anything right now, but it has the "potential" to do work in the future. You already have learned about several types of potential energy. For example, a ball at the top of a hill

Spring

has a type of potential energy, called **gravitational potential energy**, because it could roll down the hill. Batteries have potential energy in the form of stored chemical energy that is not used until the battery is connected to something. A compressed spring has potential energy that could suddenly be released if you are not careful!

CCC   **Energy and Matter**

Kinetic energy is different from potential energy. It is the movement of something. Remember riding in a car? The car's motion is kinetic energy. For some other types of kinetic energy, it might not be so obvious that something is moving. Here are some examples:

- sound or light waves moving through the air

- movement of electrons though a wire

- vibrations of molecules in a substance as it heats up

This means that sound, radiant energy, electrical energy, and thermal energy are all types of kinetic energy. They all involve something moving. Energy is transformed easily from one form into another. For example, a child at the top of a playground slide has potential energy. As the child moves down the slide, the potential energy is changed into kinetic energy.

Potential Energy	Kinetic Energy
• Chemical	• Solar
• Elastic	• Thermal
• Nuclear	• Mechanical
• Gravitational	• Electrical
• Mechanical	• Light
	• Sound

There are some other types of energy you might not think about very often. For example, chemical energy is the energy stored in chemical bonds between atoms and molecules. Nuclear energy is the energy that holds the nucleus of an atom together. Gravitational energy keeps us from floating off into space! Whether we think about it or not, each type of energy plays an important role in our daily lives.

A roller coaster will convert stored potential energy in the cars as it drags them up the first hill. What form of potential energy is it creating?

_____

When the roller coaster goes down the hill, what form of energy is it converted into?

_____

Activity 10
Observe Like a Scientist

Quick Code:
us4039s

Types of Energy

Watch the video about different types of energy. **Look** for two examples of potential energy. **Share** what happens when the potential energy is converted to kinetic energy.

Video

Types of Energy

Talk Together

Now, talk together about ways potential energy is converted to kinetic energy.

Activity 11
Observe Like a Scientist

Forms of Energy

Quick Code:
us4040s

Think about the different forms of energy around you. **Record** an example of the various forms in the chart. Then, **complete** the interactive to learn more about the different forms of energy. **Add** to your chart after you complete the Interactive.

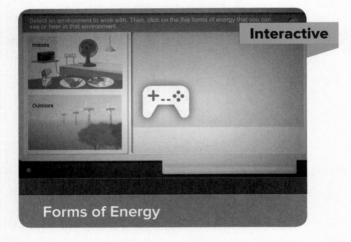

Forms of Energy

Forms of Energy	My Examples
Motion Energy	
Sound Energy	
Light Energy	
Chemical Energy	
Electrical Energy	

CCC **Energy and Matter**

Analyze Like a Scientist

Energy Transformation in Engines

Read the text. Then, **answer** the questions that follow.

Energy Transformation in Engines

You might know that cars and trucks need gasoline to run, but what is gas made of? How does it make a car move? Gasoline contains chemical energy, and a car's engine transforms that energy to power cars. Gasoline also powers trucks, boats, and many other large and small vehicles. Gasoline is one of the products that are made from oil. Oil is made from prehistoric plants and animals. The plants and animals were buried under the ground. After millions of years, the plants and animals became oil. The remains of living things that died a long time ago are called fossils. This is why oil is called a fossil fuel. The oil contains energy. The energy in oil is called chemical potential energy.

It is called potential energy because it has the potential to power many things. It is just like the food that you eat that gives you energy for your day. When the oil is collected and made into gasoline, your car can use it to produce another kind of energy. This second kind of energy is called kinetic energy. Kinetic energy is the energy of motion.

CCC **Energy and Matter**

How does this work? Look at this image.

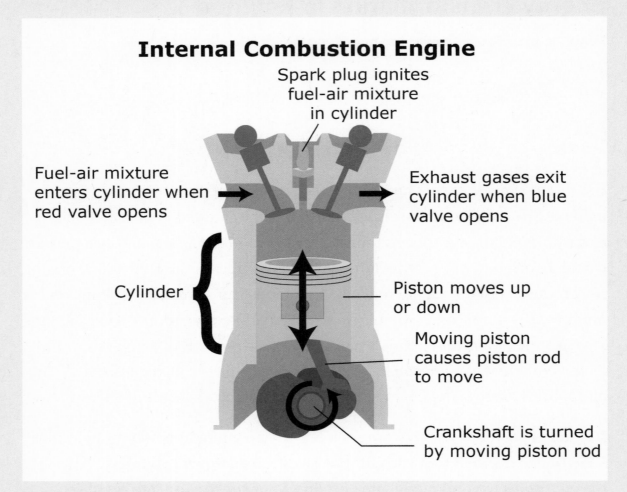

Internal Combustion Engine

Spark plug ignites
fuel-air mixture
in cylinder

Fuel-air mixture
enters cylinder when
red valve opens

Exhaust gases exit
cylinder when blue
valve opens

Cylinder

Piston moves up
or down

Moving piston
causes piston rod
to move

Crankshaft is turned
by moving piston rod

This image shows the engine inside a car or bus, which is called an internal combustion engine. An internal combustion engine can safely burn gasoline inside it. As the gasoline burns, the energy from the plants and animals from millions of years ago is changed. The energy is changed from potential energy into kinetic energy, which is motion energy. Kinetic energy is what causes the car or bus to move. One important thing to remember about energy is that it cannot be created or destroyed. Energy can only change. Potential energy can change into kinetic energy.

So, an internal combustion engine changes the chemical potential energy of the gasoline. What does the engine change it into? It changes it into kinetic energy. What does the kinetic energy do? The kinetic energy powers the car.

What does the internal combustion engine change the potential energy of gasoline into?

_____

_____

_____

_____

_____

_____

How does this compare to your body when you eat food?

_____

_____

_____

_____

_____

_____

Activity 13

Evaluate Like a Scientist

Quick Code:
us4042s

Forms of Energy

Read each scenario that uses energy. **Write** down a list of questions to help figure out which form of energy matches to the scenario. Then, use the chart that follows to **write** each situation under the correct category.

- A battery is sitting on a shelf in a store.

- The sun is shining during the day.

- The ground is cooling at night.

- A car is parked near the top of a hill.

- A battery is hooked up to a lighted lightbulb.

- A car is coasting down a long hill.

My questions:

_____

_____

_____

_____

_____

_____

_____

_____

_____

_____

CCC **Energy and Matter**

Topic: _____

Potential Energy	Kinetic Energy

Activity 14
Analyze Like a Scientist

Quick Code:
us4043s

Energy Used to Move Objects

Read the text aloud with a partner, switching for each paragraph.
Summarize the main idea of each paragraph. Then, **summarize** the main idea of the whole text.

Energy Used to Move Objects

Forces—pushes and pulls—can move objects, but where does the energy come from to produce these forces? Where does the energy needed to move a car or a jet plane come from? What about the energy needed to move a bicycle or a soccer ball?

Race Car

The fuel the car uses is most likely gasoline; that of the plane is kerosene. The bicycle and soccer ball are both human-powered. The energy used to move those comes from food—human fuel. Fuels are substances that are rich in a type of potential energy called chemical potential energy. In the engines of the car and jet, this potential energy is transformed into kinetic energy. The kinetic energy from the turning engine parts is then used to move the vehicle.

In animals, including humans, muscles use the energy that food provides to produce movement that moves limbs—arms, legs, or wings. This movement is, in turn used to move the entire animal.

The chemical potential energy in fuels is only one form of potential energy. Other forms can be used to make objects move. Objects that drop or roll down a hill are converting gravitational potential energy into kinetic energy. Roller coasters or ski jumpers use this to reach high speeds. Energy can be stored in springs and rubber bands and released, transforming their potential energy into kinetic energy. This is how toy airplanes are powered using large rubber bands.

Summary: Paragraph 1

_____

_____

_____

_____

Summary: Paragraph 2

_____

_____

_____

_____

Summary: Paragraph 3

_____

_____

_____

_____

Summary: Paragraph 4

_____

_____

_____

_____

Main Idea:

_____

_____

_____

_____

SEP **Constructing Explanations and Designing Solutions**

Activity 15
Evaluate Like a Scientist

Quick Code:
us4044s

Easy Life Tool

Think about the different forms of potential energy that make objects move. **Write** a list of tasks that would be easier to do with a tool. **Choose** one task and **draw** it, showing how energy flows.

SEP **Constructing Explanations and Designing Solutions**

Activity 16
Record Evidence Like a Scientist

Quick Code:
us4045s

Roller Coasters

Now that you have learned about energy and motion, look again at the video on roller coasters. You first saw this in Wonder.

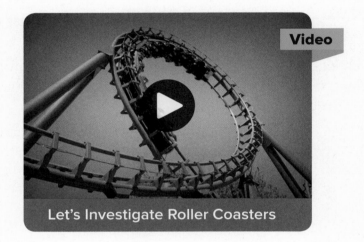

Video

Let's Investigate Roller Coasters

 Talk Together

How can you describe the roller coaster video now? How is your explanation different from before?

SEP **Constructing Explanations and Designing Solutions**

Look at the Can You Explain? question. You first read this question at the beginning of the lesson.

Can You Explain?

How do moving objects get energy?

Now, you will use your new ideas about energy and motion to answer a question.

1. **Choose** a question. You can use the Can You Explain? question or one of your own. You can also use one of the questions that you wrote at the beginning of the lesson.

My Question

2. Then, **use** the graphic organizers on the next pages to help you answer the question.

To plan your scientific explanation, first **write** your claim.

My claim:

_____

_____

_____

_____

_____

Next, look at your notes in the bubble map. **Identify** two pieces of evidence that support your claim:

Evidence 1

Evidence 2

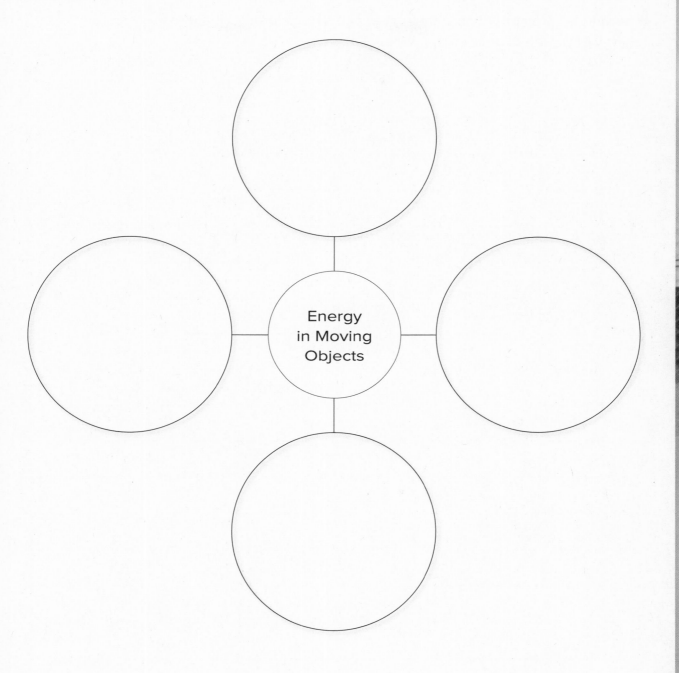

Energy
in Moving
Objects

Now, **write** your scientific explanation.

Knowing the way energy is transformed can help us predict the kinetic energy in an object because...

_____

_____

_____

_____

_____

_____

_____

_____

_____

_____

_____

_____

_____

_____

_____

_____

STEM in Action

Activity 17
Analyze Like a Scientist

Quick Code:
us4046s

Kinetic Energy and Potential Energy in Winter Sports

Read the text. **Think** about how kinetic energy and potential energy are used during ice-skating. Then, **complete** the activities that follow.

Kinetic Energy and Potential Energy in Winter Sports

Do you enjoy winter sports such as ice-skating?

Ice-skaters can do amazing things. Surya Bonaly was an ice-skater who did a backflip on the ice and landed on one skate.

Surya Bonaly

Nathan Chen might try to do seven jumps in one program!

Nathan Chen

Starr Andrews once skated to "Whip My Hair." Ice-skating can be fun to watch.

Would you like to learn to figure skate like Surya, Nathan, and Starr?

Ice rinks have ice-skating lessons. Ice-skating clubs have lessons too. Figure Skating in Harlem (FSH) is a skating club for girls in New York City. At FSH, girls from all backgrounds can learn how to ice-skate. They also learn other life lessons. For example, they learn how to be strong and confident. They discover what foods to eat to provide energy for skating and life.

Let's think about when Surya begins to skate. The potential energy in her body changes into kinetic energy. The kinetic energy and her strong leg muscles help her jump high into the air.

She is working very hard. A lot of energy is being used as well. At the top of the jump, her energy changes again. Because she is up in the air, she has potential energy. You can think of this like a basketball at the top of its arc. Gravity pulls her back down to the ice, turning her potential energy back to kinetic energy.

When does the skater have the least kinetic energy? When does she have the most kinetic energy?

_____

_____

_____

_____

More Potential Energy or More Kinetic Energy?

Look at the different pictures, and **think** about energy use. **Circle** the pictures that show more potential energy.

Activity 18

Evaluate Like a Scientist

Quick Code:
us4047s

Review:
Energy and Motion

Think about what you have read and seen in this lesson. **Write** down some core ideas you have learned. **Review** your notes with a partner. Your teacher may also have you take a practice test.

_____

_____

_____

_____

_____

_____

_____

_____

_____

_____

_____

_____

_____

_____

_____

_____

Talk Together

Think about the Science of Car Crashes video you saw in Get Started. Use your new ideas to discuss how energy conservation relates to motion.

Speed

Student Objectives

By the end of this lesson:

☐ I can apply mathematical thinking to calculate the speed of objects using standard units of measurement.

☐ I can describe how an object's change in position occurs at different rates.

☐ I can model data to show patterns in the speed of objects and use these patterns to predict future motion.

☐ I can cite evidence to explain how speed is related to the amount of kinetic energy of an object.

☐ I can explain why an object's speed can change.

Key Vocabulary

☐ speed

Quick Code:
us4048s

Activity 1
Can You Explain?

How can you measure the speed of something moving fast?

_____

_____

_____

_____

_____

_____

_____

_____

_____

_____

_____

Quick Code:
us4050s

Activity 2
Ask Questions Like a Scientist

Cheetah Speed

Watch the video. **Think** of questions you have about speed.

Quick Code:
us4051s

Video

Let's Investigate Cheetah Speed

SEP **Asking Questions and Defining Problems**

Talk Together

Now, share your questions and ideas about speed. Why is the cheetah so fast? What personal experiences have you had with high speeds?

Activity 3

Observe Like a Scientist

Catching Own Pass

Watch the video, and **look** for objects moving at different speeds.

Quick Code:
us4052s

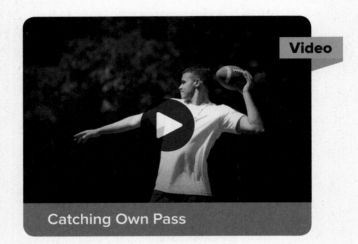

Video

Catching Own Pass

Talk Together

Now, talk together about how the speed of the player was determined.

© Discovery Education | www.discoveryeducation.com • Image: (a) yulkapopkova / E+ / Getty Images. (b) behindlens / Shutterstock.com. (c) Icon made by Freepik from www.flaticon.com

Activity 4

Observe Like a Scientist

Measuring Wind

Watch the video, and **look** for different methods of measuring speed.

Quick Code:
us4053s

Measuring Wind

Talk Together

Now, talk together about the different tools needed to measure speed.

CCC **Energy and Matter**

Activity 5

Evaluate Like a Scientist

What Do You Already Know About Speed?

Calculating Speed

Write the correct words to **complete** the mathematical formula for calculating speed.

location	acceleration	time
distance	direction	

speed = ⬜ ÷ ⬜

Slowing Down

Circle the best explanation of how to slow down an object that is moving at a certain speed.

Just wait; it will slow down by itself.

Apply a force in the direction that the object is moving.

Apply a force in the direction opposite the object's motion.

CCC **Energy and Matter**

© Discovery Education | www.discoveryeducation.com ● Image: yulkapopkova / E+ / Getty Images

What Is Speed?

Activity 6

Analyze Like a Scientist

Basics of Speed

Quick Code: us4055s

Read the text, and **look** at the image. Then, **write** and **draw** your definition of speed.

Basics of Speed

Speed is a measurement of how fast something is moving. Speed measures the distance that an object travels over time. The speed of an object is the same no matter which direction it moves. If you skate 5 meters backward every second or 5 meters forward, your speed is still 5 meters per second. Speed is displayed in units of distance over time. Therefore, to calculate an object's speed, divide the distance it travels by the time it takes to travel there. Some common units of speed are meters per second (m/sec), kilometers per hour (km/hr or kph), and miles per hour (mi/hr or mph).

Traffic Signs

To compare the speed of one object to the speed of a second object, measure the distance both objects travel in a given period of time. The object that travels the greater distance is moving at a greater speed. If one runner travels 6 kilometers in 1 hour and a second runner travels 9 kilometers in 1 hour, the second runner is moving at a greater speed.

Another way to compare speed is to see which object moves a given distance in the smallest amount of time. Imagine two cars racing 1,000 meters. The car that finishes in less time is faster. It has the greater speed.

Speed is defined as distance per unit of time. We often see speed in units of miles per hour. For instance, a car that travels 50 miles per hour is faster than a car that travels 30 miles per hour.

Speed is...

Activity 7

Observe Like a Scientist

Measuring an Object's Motion

Quick Code:
us4056s

Watch the video. As you watch, **look** for examples of speed being measured.

Measuring an Object's Motion

Talk Together

Now, talk together about ways to calculate an object's speed.

Evaluate Like a Scientist

Quick Code:
us4057s

Measuring Speed

Circle all the words and phrases that measure speed.

65 miles per hour

25° Celsius

10 kilometers

9.8 meters/second

200 kilometers per day

5° north

How Do I Calculate Speed?

Activity 9
Observe Like a Scientist

Quick Code:
us4058s

Speed and Time

Watch the video. **Look** for objects moving at different speeds.

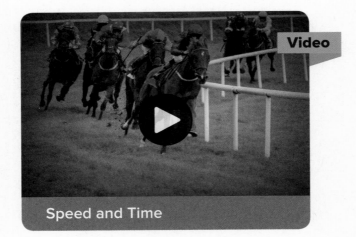

Video

Speed and Time

Talk Together

Now, talk together about how to determine which objects are moving slower and which are moving faster.

© Discovery Education | www.discoveryeducation.com • Image: (a) yulkapopkova / E+ / Getty Images. (b) gabriel12 / Shutterstock.com. (c) Icon made by Freepik from www.flaticon.com

Activity 10
Investigate Like a Scientist

Hands-On Investigation: Measuring Speed

In this investigation, you will **measure** the speed of various balls traveling down an inclined plane and then **describe** and **graph** the data. You will use your data to **calculate** average speed and will **determine** the best way to present your data.

Make a Prediction

What will you need to know to determine the speed of the balls?

_____

_____

_____

_____

What are some ways you can present the data?

_____

_____

_____

SEP	Using Mathematics and Computational Thinking
SEP	Constructing Explanations and Designing Solutions
CCC	Patterns

What materials do you need? (per group)

- Wooden board
- Tennis balls
- Golf ball
- Marbles, $\frac{5}{8}$ in.
- Ping pong balls
- Collision ball 19 mm

- Measuring tape
- Stopwatch
- Graph paper
- Calculator, solar
- Balance, double pan

What Will You Do?

1. With your group, roll a variety of balls down the inclined plane.
2. Record the data necessary to determine the speed of each ball.
3. Calculate the average speed of each ball.
4. Display your data by drawing a graph in the space provided.

Distance of Inclined Plane:

Type of Ball	Time	Average Speed

Think About the Activity

What pattern do you see in your graph? What does this pattern show you about average speed?

_____

_____

_____

_____

How were your group's results different from other groups' results? Why do you think they were different?

_____

_____

_____

_____

_____

_____

How could you change the ball's speed, other than changing the type of ball you roll down the inclined plane?

_____

_____

_____

_____

Discovery EDUCATION

Activity 11
Observe Like a Scientist

Describing Motion

Quick Code: us4060s

Perform the interactive. **Record** your data in the table. Then, **answer** the questions.

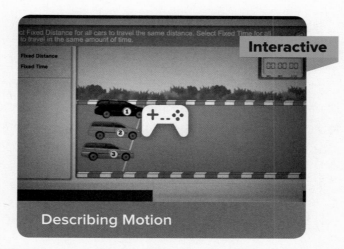

Describing Motion

| Car | Fixed Distance (30 m) | | Fixed Time (3 sec) | |
	Time Taken (sec)	Calculated Speed (m/sec)	Distance Covered (m)	Calculated Speed (m/sec)
Black				
Red				
Blue				

SEP **Constructing Explanations and Designing Solutions**

A friend says, "I'm the fastest runner in my class because I can run 4 miles without stopping." **Explain** to your friend why her statement doesn't make sense.

_____

_____

_____

_____

_____

A cyclist biked 5 kilometers in 10 minutes. She took a break and then biked another 5 kilometers in 14 minutes. Did her speed increase or decrease after she took her break?

_____

_____

_____

A runner finished the 100-meter race in 20 seconds. What was his speed in meters per second?

_____

_____

Activity 12

Analyze Like a Scientist

Calculating Speed

Quick Code:
us4061s

Read the text. Then, in the space provided, **write** a speed problem for your partner to solve.

Calculating Speed

Let's use some time and distance data to calculate the speed of two different colored cars. A yellow car moves 100 meters in 10 seconds. A green car moves 15 meters in 5 seconds. What are the speeds of the two cars? Which car is going faster? First, we will calculate the speed of the yellow car:

100 m in 10 sec=**100/10** m/sec=10 m/sec, or 10 meters per second

Next, we will calculate the speed of the green car:

15 m in 5 sec=**15/5** m/sec=3 m/sec, or 3 meters per second

So, in every second, the yellow car travels 10 meters and the green car travels 3 meters. The yellow car is faster. It is more than three times as fast as the green car.

Your Speed Problem

_____

_____

SEP **Using Mathematics and Computational Thinking**

Activity 13

Think Like a Scientist

Quick Code:
us4062s

Practicing Speed Calculations

In this activity, you will calculate the speed and distance in different situations.

What Will You Do?

Calculate the speed in each situation described.

Manu has begun walking to school. It takes her 40 minutes to walk the 5 km from her home to the school. How fast is Manu walking?

_____

_____

_____

Larissa is taking a train to visit her grandmother, who lives 180 km away. If the trip takes 3 hours, how fast is the train moving?

_____

_____

_____

Jorge wants to beat his old time in his town's annual bicycle race. His new goal is to ride the 38 km race in 2 hours. How fast does Jorge need to ride to achieve this goal?

_____

_____

_____

SEP **Using Mathematics and Computational Thinking**

Chris is trying to build up his endurance so he can run a marathon. Currently, he runs at a speed of 12 km/hr. If he maintains this speed for half an hour, how far will Chris run?

_____

_____

_____

_____

Patty spent Saturday riding her bike to visit friends. In the morning, she rode for 30 minutes at a speed of 18 km/hr. In the afternoon, she rode for 40 minutes at a speed of 17 km/hr. In the evening, she rode home for 50 minutes at a speed of 18 km/hr. What is the total distance Patty rode?

_____

_____

_____

_____

Activity 14

Evaluate Like a Scientist

Quick Code:
us4063s

Comparing Speeds

Write numbers 1 through 6 to **order** the objects from fastest (1) to slowest (6).

_____ bird that flies 385 km in 1 hour

_____ airplane that flies 1,600 km in 2 hours

_____ race boat that travels 450 km in 3 hours

_____ race car that moves 1,400 km in 4 hours

_____ train that moves 1,250 km in 5 hours

_____ sailfish that travels 660 km in 6 hours

SEP **Asking Questions and Defining Problems**

What Is the Relationship between Speed and Kinetic Energy?

Activity 15

Investigate Like a Scientist

Quick Code: us4064s

Hands-On Investigation: Hit the Slopes!

In this investigation, you will use model trucks to measure the speed and kinetic energy of objects moving down inclines at various angles. You will measure the distance a paper cup moves when your truck rolls down the incline at each angle and into the cup.

Make a Prediction

How do you think kinetic energy will change with the angle of the inclined plane?

_____

_____

_____

How will the cup measure kinetic energy?

_____

_____

_____

SEP Developing and Using Models

What materials do you need? (per group)

- Toy trucks
- Cardboard paper towel tube
- Paper cup, 360 mL
- Scissors
- Several books

- Protractor
- Metric ruler
- Removable sticky note flags
- Stopwatch

HANDS-ON INVESTIGATION

What Will You Do?

1. With your partner, measure and record the angle of your tube.

2. Roll your truck down the tube, and record how long it takes to travel to the end of the tube.

3. Change the angle, and repeat the steps.

4. Now, repeat each angle, but place a cup at the bottom of the tube.

5. Measure the distance the cup moves after each time the truck rolls into it.

Angle of Tube	Elapsed Time	Distance the Cup Moved

Think About the Activity

What happened to the speed of the truck when the incline increased?

_____

_____

How did the results of the speed test compare to the results of the kinetic energy test?

_____

_____

_____

What conclusion can you draw about the relationship between speed and kinetic energy, based on this experiment?

_____

_____

_____

_____

Activity 16
Analyze Like a Scientist

Changing Speed

Quick Code:
us4065s

Read the text. As you read, **highlight** the key ideas and supporting details.

Changing Speed

If you want an object to move faster, you must give it more kinetic energy. If you want it to go slower, you must reduce its kinetic energy. You have learned that forces make things move. When a force is used to push an object, the speed of the object depends upon the force used. The more force applied to an object, the faster it goes. The faster it goes, the more kinetic energy it has. Let's think about how this works in a car.

If a driver wants a car to go faster, she presses the gas pedal. This sends more gasoline into the engine. This allows the engine to convert more potential energy in the gas into kinetic energy in the engine. This kinetic energy provides the force that turns the wheels faster, and the car speeds up.

Changing Speed

What if the driver wants the car to slow down? If she takes her foot off the gas, the car will slowly roll to a stop because of friction. If she presses the gas pedal less, the car will also slow down until the car reaches a slower speed. At that speed, enough gas is going to the engine to maintain the car's kinetic energy. What if she wants to stop the car quickly? She can increase the friction using the brakes. The brakes rub against the inside of the wheels, and this friction quickly slows the car down.

Activity 17

Observe Like a Scientist

Quick Code: us4066s

RC Racing Cars

Perform the interactive. **Record** your data in the table for both levels.

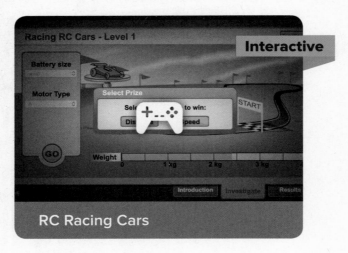

RC Racing Cars

Level 1

Test Number	Battery Size	Type of Motor	Number of Laps	Time for 5 Laps

© Discovery Education | www.discoveryeducation.com • Image: yulkapopkova / E+ / Getty Images

Discovery EDUCATION

Level 2

Test Number	Battery Size	Type of Motor	Size of Wheel	Total Cost of Car	Number of Laps	Time for 5 Laps

SEP **Planning and Carrying Out Investigations**

CCC **Energy and Matter**

Activity 18

Evaluate Like a Scientist

Train Race and Going for a Drive

Quick Code:
us4067s

Train Race

Read the text, and **answer** the question.

> Ichiro loves model trains. He wants to get one that is faster than the one he has now. The train catalog gives the speed for a new train: it travels 4 meters every 8 seconds. Ichiro tested his old train on his 3-meter track so he could compare it to the new train in the catalog. His old train traveled 3 meters in 12 seconds.

Should Ichiro buy the new train? **Explain** your reasoning using calculations of speed from the data.

_____

_____

_____

_____

_____

CCC **Energy and Matter**

Going for a Drive

Ayana's mother is driving to the dentist. The dentist's office is 38 kilometers from Ayana's home. **Look** at the graph, and then **answer** the questions.

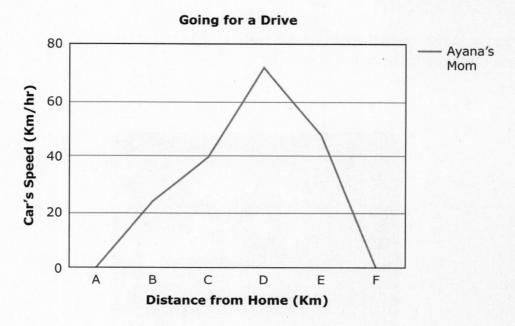

At what point on the graph is Ayana's mother going the fastest?

_____

At what point on the graph does the car have the most kinetic energy, and why?

_____

_____

At what point(s) on the graph does the car have the most potential energy, and why?

_____

Activity 19

Record Evidence Like a Scientist

Quick Code:
us4068s

Cheetah Speed

Now that you have learned about speed, **watch** the video of the cheetah again. You first saw this in Wonder.

Video

Let's Investigate Cheetah Speed

Talk Together

How can you describe cheetah speed now? How is your explanation different from before?

SEP **Constructing Explanations and Designing Solutions**

© Discovery Education | www.discoveryeducation.com ● Image: (a) yulkapopkova / E+ / Getty Images, (b) JonathanC Photography / Shutterstock.com, (c) Icon made by Freepik from www.flaticon.com

Look at the Can You Explain? question. You first read this question at the beginning of the lesson.

Can You Explain?

How can you measure the speed of something moving fast?

Now, you will use your new ideas about speed to answer a question.

1. **Choose** a question. You can use the Can You Explain? question or one of your own. You can also use one of the questions that you wrote at the beginning of the lesson.

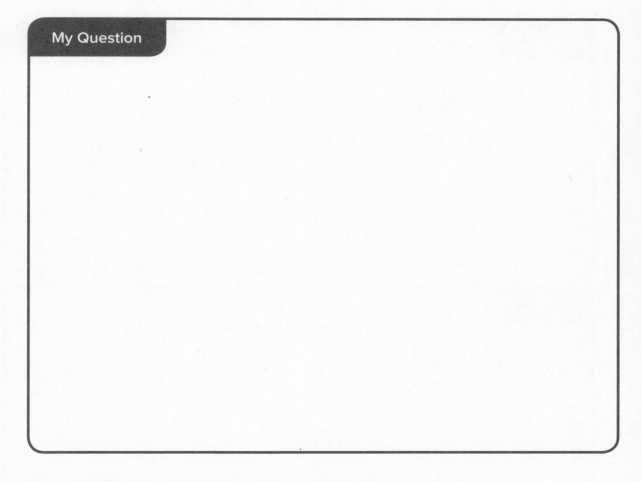

My Question

2. Then, **use** the graphic organizers on the next pages to help you answer the question.

To plan your scientific explanation, first **write** your claim. You claim is a one-sentence answer to the question you investigated. It answers: What can you conclude? It should not start with *yes* or *no*.

My claim:

_____

_____

_____

_____

Data should support your claim. Leave out information that does not support your claim. **List** data that supports your claim.

> **Data 1**
>
>

> **Data 2**
>
>

Finally, **explain** your reasoning. Reasoning ties together the claim and the evidence. Reasoning shows how or why the data count as evidence to support the claim.

Topic: _____

Evidence	How Supports Claim

Now, **write** your scientific explanation.

_____

_____

_____

_____

_____

_____

_____

_____

_____

_____

_____

_____

_____

_____

_____

_____

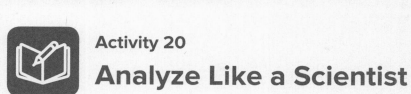

STEM in Action

Quick Code:
us4069s

Activity 20

Analyze Like a Scientist

Solar Vehicles

Read the text. As you read, **complete** the graphic organizer. Then, **complete** the activity.

Solar Vehicles

Most cars run on gasoline. This requires trips to the gas station and contributes to climate change. More and more vehicles are running on electricity. Electric vehicles have batteries that must be charged. Can you imagine a car that never has to stop for gas or a charge?

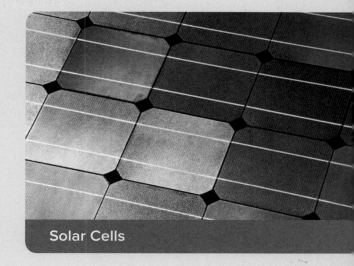

Solar Cells

Mechanical engineers are designing vehicles that run on nothing but the sun. There are, of course, some difficulties. At this time, the amount of energy we can capture from the sun is not as great as the amount of energy we get from gasoline or an electric battery. How can mechanical engineers make solar vehicles drive as quickly as conventional vehicles? Among other things, they reduce the weight of the vehicle and they make it incredibly aerodynamic.

Pros of Using This Car	Cons of Using This Car

Solar Vehicle

The solar vehicle is so lightweight that most instruments are excluded. Without a speedometer, how can we know the speed of the solar vehicle? In the following activity, you will design a way to calculate the solar vehicle's speed.

The fastest solar vehicle can go slightly more than 55 miles per hour. Calculating that speed can be challenging because most of the solar vehicle races take place in remote locations, and in most cases, the solar vehicles do not have speedometers. Imagine that you have been given the task of calculating the speed of the solar vehicle. How would you do it?

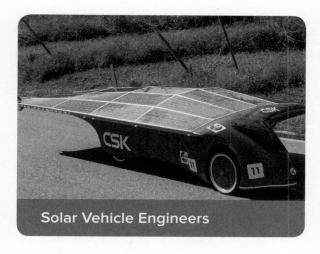

Solar Vehicle Engineers

_____

_____

_____

_____

_____

_____

_____

Activity 21

Evaluate Like a Scientist

Quick Code:
us4070s

Review: Speed

Think about what you have read and seen in this lesson. **Write** down some core ideas you have learned. **Review** your notes with a partner. Your teacher may also have you take a practice test.

_____

_____

_____

_____

_____

_____

_____

_____

_____

_____

_____

_____

_____

SEP Obtaining, Evaluating, and Communicating Information

Talk Together

Think about the Science of Car Crashes video you saw in Get Started. Use your new ideas to discuss speed and kinetic energy, how they are related and how they are calculated.

Energy and Collisions

Student Objectives

By the end of this lesson:

☐ I can analyze and interpret data to describe how the speed and mass of objects relate to changes observed in a collision.

☐ I can construct an explanation based on evidence and logical reasoning to describe energy transfer in a collision.

☐ I can apply mathematical thinking to organize data to represent patterns related to mass, speed, and the energy of objects.

Key Vocabulary

☐ acceleration   ☐ mass

☐ collision

Quick Code:
us4071s

Activity 1

Can You Explain?

What happens to objects when they collide with another object?

_____

_____

_____

_____

_____

_____

_____

_____

_____

_____

Quick Code:
us4073s

DISCOVERY EDUCATION

Activity 2

Ask Questions Like a Scientist

Hitting a Home Run

Look at the image. Then, **discuss** the questions.

Quick Code:
us4074s

Let's Investigate Hitting a Home Run

© Discovery Education | www.discoveryeducation.com ● Images: (a) mevans / E+ / Getty Images, (b) John Giustina / Iconica / Getty Images

SEP Asking Questions and Defining Problems

Discuss with Your Class

What happens when two objects hit each other? Imagine you are watching a softball game. The player stands with her bat and wields it as the ball approaches her at high speed. The bat makes contact with the ball. What happens at that moment of contact? What do your senses observe? What would the player feel?

Activity 3

Observe Like a Scientist

Quick Code:
us4075s

Watching Objects Collide

Watch the videos. As you watch, **look for** objects changing motion as they collide with one another.

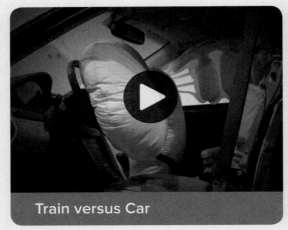

Train versus Car

Colliding Worlds

Videos

© Discovery Education | www.discoveryeducation.com • Images: (a) mevans / E+ / Getty Images, (b) Sony Ho / Shutterstock.com, (c) Vadim Sadovski / Shutterstock.com, (d) Icon made by Freepik from www.flaticon.com.

🗨️ Talk Together

Now, talk together about the factors that cause objects to change their motion.

CCC **Cause and Effect**

Activity 4

Evaluate Like a Scientist

Quick Code:
us4076s

What Do You Already Know About Energy and Collisions?

Cycling Collision

A bicyclist is riding down the road at a high speed. The bicyclist does not notice an empty metal trash can in her path. The cyclist hits the can. **Circle** the effects you predict the collision will have.

The trash can will move.

The trash can will gain energy.

The speed of the bicycle will increase.

Some of the energy of the bicycle's motion will be destroyed.

Some of the energy of the bicycle's motion will be changed into sound.

CCC **Cause and Effect**

Activity 5

Analyze Like a Scientist

Energy and Collisions

Quick Code:
us4077s

Read the text. Then, **complete** the activity.

Energy and Collisions

When two things bump, or crash, into each other, we can say a **collision** has taken place. When this happens, the two objects exchange energy. Energy changes also occur. Think about this: If you are running down the street without looking, and you run into a sign, what happens? The chances are you will stop moving, perhaps bounce off, and get hurt. The sign, if it is made of iron, may wobble a bit and rattle. When you hit the sign, you would stop moving forward. What happened to your kinetic energy? What energy changes were taking place here? How would things be different if you were walking? What could have happened if you were running faster?

Children Running

Now, **draw** a two-framed comic strip showing the before and after of a collision. Underneath, **write** a description of the changes in kinetic energy of the objects that collide.

How Does the Speed of an Object Affect What Happens in a Collision?

Activity 6

Analyze Like a Scientist

The Effect of Speed on Collisions

Quick Code: us4078s

Read the text. As you read, **highlight** information in the text that supports the patterns you saw in the data from the activity in the previous concept. In that concept, you used model cars to measure the speed and kinetic energy of objects moving down inclines of various angles.

The Effect of Speed on Collisions

The amount of kinetic energy an object has depends upon its speed. The faster an object travels, the more energy it has. When a speeding object hits another object, it transfers some of its energy to it. The faster the object, the more energy it transfers. Some of this energy may be in the form of heat, light, or sound. This is why a fast softball makes more sound as it hits the bat than a slow one. Because of their extra energy, fast-moving objects can do much more damage than slow ones. When they hit an object, they exert more force. This force can smash a car fender or, in some cases, damage the car beyond repair.

SEP Asking Questions and Defining Problems

CCC Patterns

Accident

Recall that if a car doubles its speed, its kinetic energy increases fourfold. All this energy will result in a large force being exerted in an accident. This is one reason why driving fast is so hazardous. If two cars drive headlong into one another, then the forces exerted in the accident depend upon the combined speed of both cars. Damage would be much more severe. What do you think would happen if two cars traveling at different speeds in the same direction collided? How would the forces in this rear-end collision compare to those in a headlong collision?

Activity 7

Investigate Like a Scientist

Hands-On Investigation: Speed and Collisions

Quick Code:
us4079s

In this investigation, you will use marbles and tubes to measure the speed and kinetic energy of objects. These objects will move down inclines of various angles.

Make a Prediction

How do you think the angle of the incline will affect the kinetic energy of an object?

_____

_____

_____

How are speed and kinetic energy related?

_____

_____

_____

SEP Engaging in Argument from Evidence

What materials do you need? (per group)

- Marbles, 1 in.
- Cardboard paper towel tube
- Modeling clay, red
- Several books
- Protractor

- Metric ruler
- Stopwatch
- Flour (optional)
- Water (optional)

What Will You Do?

1. With your partner, determine the best way to record your data in the space.

2. Measure and record the angle of the incline.

3. Measure and record the time it takes the marble to roll down the incline.

4. Increase the angle and repeat the first two steps, for a total of three trials.

5. Place a piece of modeling clay at the end of the incline.

6. For each of the three incline angles you measured earlier, roll a marble down the incline so it hits the modeling clay.

7. Measure and record the deepest part of the indent the marble made in the clay.

Number of Books	Depth of Dent

Think About the Activity

How did the results of the speed test compare with the results of the kinetic energy test?

_____

_____

_____

How do the results from this experiment compare with the results from the tests you did in Hit the Slopes? How are they different?

_____

_____

_____

_____

_____

_____

What can you conclude about the relationship between speed and kinetic energy based on this experiment?

_____

_____

_____

_____

What does the depth of the dents tell you about what happens to vehicles in a real-world collision?

_____

_____

_____

Activity 8

Evaluate Like a Scientist

Micrometeorites

Quick Code: us4080s

Read the statement that follows. Then, **answer** the question.

"In space, tiny fragments of rock, called micrometeorites, travel at very high speeds. These micrometeorites are about the size of a grain of sand. If one were to hit an astronaut, it could puncture the astronaut's suit. This could kill the astronaut."

Explain in terms of kinetic energy how such a tiny object could cause so much harm.

_____

_____

_____

_____

_____

_____

_____

_____

_____

_____

_____

CCC **Energy and Matter**

How Does the Mass of an Object Affect What Happens in a Collision?

Activity 9

Investigate Like a Scientist

Quick Code:
us4081s

Hands-On Investigation: Mass in Collisions

In this investigation, you will use marbles of different masses moving down inclines to measure their speed and kinetic energy. You will measure the distance a paper cup moves when each marble rolls down the incline into the paper cup. You will also measure the dent made in a block of modeling compound when each marble rolls down the incline of each angle and into the compound.

Make a Prediction

How are mass and speed related?

_____

_____

_____

How are mass and kinetic energy related?

_____

_____

_____

SEP **Engaging in Argument from Evidence**

What materials do you need? (per group)

- Marble, 1 in.

- Collision ball, 16 mm

- Collision ball, 19 mm

- Cardboard paper towel tube

- Paper cup, 360 mL

- Scissors

- Modeling clay, red

- Several large books

- Metric ruler

- Removable sticky note flags

- Stopwatch

- Balance, double pan

What Will You Do?

1. With your partner, decide how you will record your data in the area that follows.

2. Choose an angle, and set your incline to that same angle for the entire investigation.

3. Roll marbles of different mass down the incline, and measure and record their speed.

4. Create a plan for measuring the kinetic energy of the marbles of different mass to find how their kinetic energy changes as their mass changes.

5. Follow your plan, and record your data.

6. Create a plan for measuring the kinetic energy of the marbles of different mass to find how their kinetic energy changes in collisions as their mass changes.

7. Follow your plan, and record your data.

Plan and Data for Measuring Speed

Discovery EDUCATION

Plan and Data for Measuring Kinetic Energy

Plan and Data for Measuring Kinetic Energy in Collisions

Discovery EDUCATION

Think About the Activity

What happened to the speed of the marble when its mass increased?

_____

_____

_____

_____

_____

How did the results of the speed test compare to the results of the kinetic energy test?

_____

_____

_____

_____

How do the results from this experiment compare to the results from the tests you did in Hit the Slopes! and Speed and Collisions? How are they different?

_____

_____

_____

_____

_____

What do you think would happen if you used a marble with greater mass than in your previous experiments?

_____

_____

_____

_____

_____

What do your results tell you about vehicle collisions in the real world?

_____

_____

_____

_____

_____

Activity 10
Analyze Like a Scientist

The Effect of Mass on Collisions

Quick Code:
us4082s

Read the text. Then, **choose** two conversation starters to **discuss** with your classmates.

The Effect of Mass on Collisions

Why do you need a bigger engine to move a semi than you do to move a car? The difference has to do with the **mass** of each vehicle. A semi has a much larger mass than a car. As each vehicle moves faster, the energy from the fuel its engine uses is converted into kinetic energy.

Comparing Trucks

The bigger the mass of the vehicle, the more fuel it consumes, and the more kinetic energy it gains. A semi traveling at the same speed as a car has more kinetic energy. If the mass of an object doubles, its kinetic energy at a certain speed also doubles. So, a 1-ton truck has half the kinetic energy of a 2-ton truck traveling at the same speed.

CCC **Scale, Proportion, and Quantity**

The Effect of Mass on Collisions *cont'd*

This is why when a big vehicle collides with something, it causes a lot more damage than a small one traveling at the same speed. A pedestrian hit by a bike traveling at 50 kilometers per hour would almost certainly survive. A car traveling at that speed would probably kill a pedestrian. Being hit by a big truck would definitely be fatal.

Now, **choose** two of the conversation starters from the chart. **Discuss** what you have read.

Question	Clarity	Connect
I don't get this part...	Let me explain...	This reminds me of...
What if...	No, I think it means...	The differences are...
Predict	**Comment**	**Explain**
I wonder if...	This is confusing because...	The basic idea is...
I think that...	This is hard because...	My understanding is...

Activity 11

Analyze Like a Scientist

Quick Code:
us4083s

Energy Conversions during a Collision

Read the text. As you read, **highlight** all the forms of energy to which kinetic energy is transformed when the balls strike one another.

Energy Conversions during a Collision

From what you have observed, you know that when objects collide, energy changes and transfers take place. The amount of energy depends on the kinetic energy of the objects and the direction in which they are traveling. Their kinetic energy is determined by both their speed and their mass. What happens to all this kinetic energy when objects collide?

None of the energy disappears. In a collision, energy in equals energy out. Energy is conserved in a collision. We can model collisions using a simple device called a Newton's cradle. In a Newton's cradle, most of the energy is transferred to other balls, which is why the same number of balls move on one side of the cradle as on the other.

CCC **Energy and Matter**

Newton's Cradle

You can hear that some energy is lost as sound. Some is lost as friction between the string and other parts, as the balls move. The balls lose a little energy as they pass through the air. If you leave the cradle long enough, after lots of collisions, the moving balls lose their kinetic energy and stop.

Activity 12
Investigate Like a Scientist

Quick Code:
us4084s

Hands-On Investigation: What Happens to Energy in Collisions?

In this investigation, you will observe a wide variety of collisions and apply what you have learned to determine what is happening to the energy in terms of transfer and change in each situation.

Make a Prediction

How is the sound of a collision related to energy?

_____

_____

_____

_____

_____

When a rolling ball hits a stationary ball, what will happen to the stationary ball?

_____

_____

_____

_____

_____

CCC **Patterns**

What materials do you need? (per group)

- Pull back car
- Toy trucks
- Toy trucks, pull back
- Cardboard paper towel tube
- Several books
- Scissors
- Removable sticky note flags
- Brick
- Foam block

- Sand
- Washers
- Rubber bands
- Large rubber band
- Shoebox
- Plastic zipper bags
- Water, 1 cup

HANDS-ON INVESTIGATION

What Will You Do?

1. With your partner, decide how you will record your data in the space that follows.

2. Build your incline with a tube and several books. Mark the spot 5 cm from the end of your tube.

3. Place a brick at the marked spot, and release a car down the incline.

4. Make and record observations and any measurements you think necessary.

5. Repeat the trial with a foam brick, 2 cups of sand, a baggie of water, a rubber band barrier, a second car facing the same direction as the moving car, a second car turned sideways to create a side-impact collision, two cars lined up, and two cars with mass added to first one and then the other.

6. Repeat the trial with any other situations you can imagine and simulate.

7. Be sure to record all data.

Data

Think About the Activity

What patterns do you notice in the behavior of the rolling car when it collides with different objects?

_____

_____

_____

_____

_____

How can you interpret these patterns to understand what happens to the car's kinetic energy in a collision?

_____

_____

_____

_____

_____

Evaluate Like a Scientist

Quick Code:
us4085s

Kinetic Energy of Vehicles

The images show five vehicles, along with their masses and speeds.
Write 1 through 5 on the lines under the vehicles to indicate the amount
of their kinetic energy, with 1 being the least and 5 being the most.

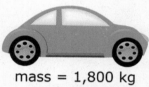

mass = 1,800 kg
speed = 15 m/s

_____

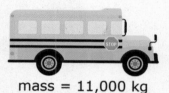

mass = 11,000 kg
speed = 0 m/s

_____

mass = 37,000 kg
speed = 15 m/s

_____

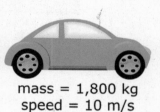

mass = 1,800 kg
speed = 10 m/s

_____

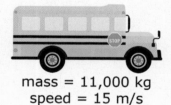

mass = 11,000 kg
speed = 15 m/s

_____

CCC **Energy and Matter**

Activity 14

Record Evidence Like a Scientist

Quick Code:
us4086s

Hitting a Home Run

Now that you have learned about energy and collisions, look again at the image of hitting a home run. You first saw this in Wonder.

Let's Investigate Hitting a Home Run

  Talk Together

How can you describe hitting a home run now? How is your explanation different from before?

SEP   **Constructing Explanations and Designing Solutions**

Look at the Can You Explain? question. You first read this question at the beginning of the lesson.

Can You Explain?

What happens to objects when they collide with another object?

Now, you will use your new ideas about hitting a home run to answer a question.

1. **Choose** a question. You can use the Can You Explain? question or one of your own. You can also use one of the questions that you wrote at the beginning of the lesson.

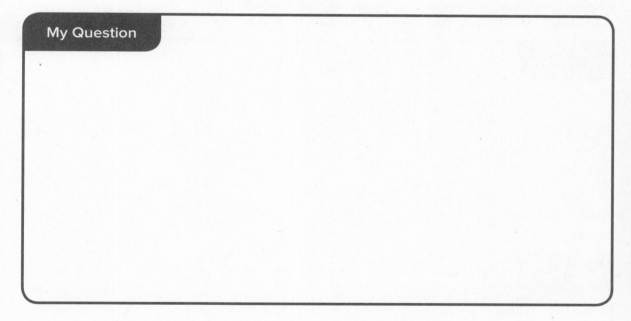

My Question

2. Then, **use** the graphic organizers on the next pages to help you answer the question.

To plan your scientific explanation, first **write** your claim. You claim is a one-sentence answer to the question you investigated. It answers: What can you conclude? It should not start with *yes* or *no*.

My claim:

_____

_____

_____

_____

_____

Data should support your claim. Leave out information that does not support your claim. **List** data that supports your claim.

Data 1

Data 2

© Discovery Education | www.discoveryeducation.com ● Image: mevans / E+ / Getty Images

Explain your reasoning. Reasoning ties together the claim and the evidence. Reasoning shows how or why the data count as evidence to support the claim.

Evidence	How Supports Claim

Now, **write** your scientific explanation.

_____

_____

_____

_____

_____

_____

_____

_____

_____

_____

_____

_____

_____

_____

_____

_____

_____

_____

Discovery
EDUCATION

S T E M in Action

Activity 15

Analyze Like a Scientist

Crash Investigator

Read the text. As you read, **highlight** the different measurements that a crash investigator needs to take.

Quick Code: us4087s

Crash Investigator

Crash investigators are police officers. They see a car crash as a puzzle. To solve the puzzle, they use Newton's laws of motion. Newton's first law states that an object at rest or in motion tends to stay at rest or in motion until acted upon by a force. Newton's second law is described by an equation: force is equal to mass times **acceleration**.

Car Crash

Crash investigators use crashed cars to learn what happened in a crash. They must find out what caused a crash. The investigator must know who was driving each car and decide who was at fault. If a crime took place, the investigator's job is to discover who committed it. By applying Newton's laws, the investigator can learn about the crash. The investigator's first task is to measure things at the scene of the accident. He or she measures damage to the cars and where the cars ended up after the crash. Sometimes, he or she may not take measurements at the scene directly. Instead, photos and videos provide needed information of the crash scene. By looking at photos in detail, the investigator can learn about the crash without blocking the road. The cars are stored to allow the damage to be closely inspected.

To apply Newton's second law, crash investigators need information. They need to know the force that acted on a vehicle and the mass of the vehicle. They measure the mass directly using a scale. To figure out the force, they use reference materials. These reference materials are measurements that the car manufacturers supply. Manufacturers crash cars under controlled conditions. They install devices that measure the forces directly. The damage to the vehicle changes with changing force. Crash investigators compare the cars from the crash to the data the manufacturers supply. This comparison helps them know how much force was involved in the crash.

Crash Site: T-Bone, Head-On, and Rear End

Read the scenarios and **complete** the activities.

A crash investigator is investigating a crash site. She has drawn this diagram of the cars just before the collision. The red car drove through the intersection from a stop, while the blue car continued in a straight line. The blue car hit the red car. **Draw** an arrow to show the direction the red car traveled after the collision. Assume the cars have about equal mass.

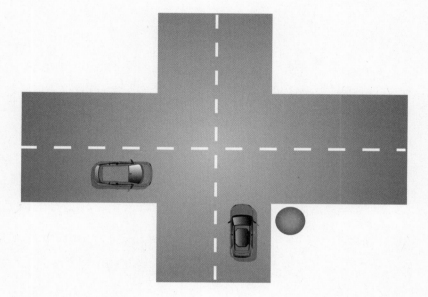

A crash investigator is investigating another crash site. She has drawn this diagram of the cars just before the collision. The red car is driving toward the intersection legally. The blue car is driving in the wrong lane. The cars are heading toward each other. The investigator's data show that the cars hit head-on. **Draw** an arrow to show the direction the red car traveled after the collision. The blue car was speeding, while the red car was below the speed limit. Assume the cars have about equal mass.

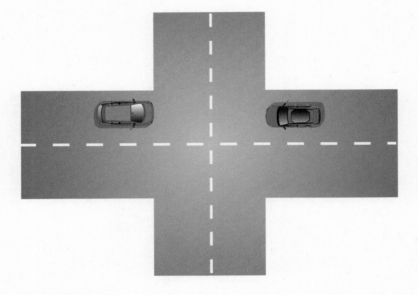

A crash investigator is investigating another crash site. She has drawn this diagram of the cars just before the collision. The red car was stopped at the stop sign. The blue car drove in a straight line. The blue car hit the red car from behind. **Draw** an arrow to show the direction the red car traveled after the collision. Assume the cars have about equal mass.

Activity 16

Evaluate Like a Scientist

Quick Code: us4088s

Review: Energy and Collisions

Think about what you have read and seen in this lesson. **Write** down some core ideas you have learned. **Review** your notes with a partner. Your teacher may also have you take a practice test.

_____

_____

_____

_____

_____

_____

_____

_____

_____

_____

_____

_____

_____

_____

_____

SEP **Obtaining, Evaluating, and Communicating Information**

© Discovery Education | www.discoveryeducation.com ● Image: mevans / E+ / Getty Images

Talk Together

Think about the Science of Car Crashes video you saw in Get Started. Use your new ideas to discuss the energy that is present in collisions, how it is transferred, and how this affects car collisions.

Solve Problems Like a Scientist

Unit Project: Self-Driving Cars

Quick Code: us4090s

Many companies are developing self-driving cars. These cars will be able to move on the roads without human drivers. To do this, the cars need many new technologies. They need sensors to detect their surroundings and computers to react to road conditions.

Collision

SEP Developing and Using Models

SEP Engaging in Argument from Evidence

These new technologies will change the way we live. These changes can be positive or negative. What are the benefits of self-driving cars? What are the risks? **Watch** the video, and **complete** the activities that follow.

Driverless Vehicle Technology

Pro or Con?

Self-driving technology will cause many changes. **Decide** whether each of the following statements is a benefit or a drawback of self-driving cars. **Mark** the benefits with a "B" and the drawbacks with a "D."

_____ Everyone can read in the car.

_____ People will not drive drunk.

_____ Cars will drive more efficiently.

_____ People will not be aware of their surroundings.

_____ Cars may not know how to react to unexpected changes.

_____ Cars will always follow the speed limit.

Model

While it is on the road, a self-driving car sometimes needs to decide whether its passenger or a pedestrian is more important. For example, **think** about a situation in which a pedestrian is slowly jaywalking across the street. The car's passenger is in a rush to arrive at her destination. **Write** what you think the car should do in this situation. **Explain** your reasoning.

_____

_____

_____

_____

_____

_____

_____

_____

_____

_____

_____

_____

New Law

What is one new law that needs to be implemented to help self-driving cars move on the road? **Write** a short description of your law. **Explain** the problem the law will solve and how the law will solve the problem.

_____

_____

_____

_____

_____

_____

_____

_____

_____

_____

Grade 4 Resources

- **Bubble Map**
- **Safety in the Science Classroom**
- **Vocabulary Flash Cards**
- **Glossary**
- **Index**

Name _____

Bubble Map

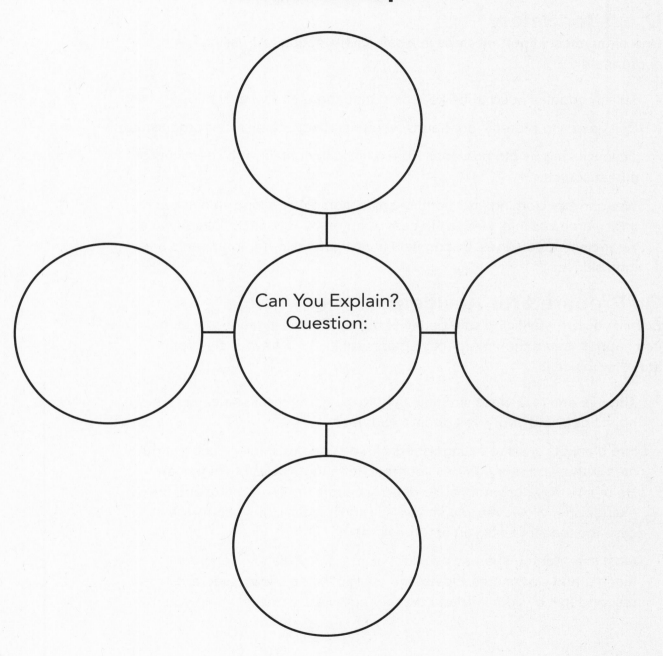

Can You Explain?
Question:

Safety in the Science Classroom

Following common safety practices is the first rule of any laboratory or field scientific investigation.

Dress for Safety

One of the most important steps in a safe investigation is dressing appropriately.

- Splash goggles need to be kept on during the entire investigation.

- Use gloves to protect your hands when handling chemicals or organisms.

- Tie back long hair to prevent it from coming in contact with chemicals or a heat source.

- Wear proper clothing and clothing protection. Roll up long sleeves, and if they are available, wear a lab coat or apron over your clothes. Always wear close toed shoes. During field investigations, wear long pants and long sleeves.

Be Prepared for Accidents

Even if you are practicing safe behavior during an investigation, accidents can happen. Learn the emergency equipment location in your classroom and how to use it.

- The eye and face wash station can help if a harmful substance or foreign object gets into your eyes or onto your face.

- Fire blankets and fire extinguishers can be used to smother and put out fires in the laboratory. Talk to your teacher about fire safety in the lab. He or she may not want you to directly handle the fire blanket and fire extinguisher. However, you should still know where these items are in case the teacher asks you to retrieve them.

- Most importantly, when an accident occurs, immediately alert your teacher and classmates. Do not try to keep the accident a secret or respond to it by yourself. Your teacher and classmates can help you.

Practice Safe Behavior

There are many ways to stay safe during a scientific investigation. You should always use safe and appropriate behavior before, during, and after your investigation.

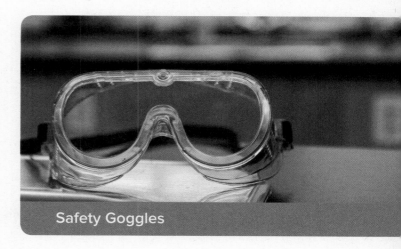
Safety Goggles

- Read the all of the steps of the procedure before beginning your investigation. Make sure you understand all the steps. Ask your teacher for help if you do not understand any part of the procedure.

- Gather all your materials and keep your workstation neat and organized. Label any chemicals you are using.

- During the investigation, be sure to follow the steps of the procedure exactly. Use only directions and materials that have been approved by your teacher.

- Eating and drinking are not allowed during an investigation. If asked to observe the odor of a substance, do so using the correct procedure known as wafting, in which you cup your hand over the container holding the substance and gently wave enough air toward your face to make sense of the smell.

- When performing investigations, stay focused on the steps of the procedure and your behavior during the investigation. During investigations, there are many materials and equipment that can cause injuries.

- Treat animals and plants with respect during an investigation.

- After the investigation is over, appropriately dispose of any chemicals or other materials that you have used. Ask your teacher if you are unsure of how to dispose of anything.

- Make sure that you have returned any extra materials and pieces of equipment to the correct storage space.

- Leave your workstation clean and neat. Wash your hands thoroughly.

acceleration

Image: Comstock / Stockbyte / Getty Images

to increase speed

chemical energy

Image: Discovery Education

the energy that is stored in the bonds between atoms

collide

Image: Pixabay

the moment where two objects hit or make contact in a forceful way

conservation of energy

Image: zsolt_uveges / Shutterstock.com

Energy cannot be created out of nothing, nor can it be completely destroyed. Energy can only change its form.

energy

Image: Paul Fuqua

the ability to do work or cause change; the ability to move an object some distance

force

Image: NASA

a pull or push that is applied to an object

friction

Image: Corepics VOF / Shutterstock.com

a force which stops motion

gravitational potential energy

Image: Adwo / Shutterstock.com

energy stored due to position in a gravitational field; commonly due to Earth's gravity

gravity

Image: Mauricio Graiki / Shutterstock.com

the force that pulls an object toward the center of Earth

heat

Image: U.S.D.A. Forest Service

the transfer of thermal energy

kinetic energy

Image: U.S. Department of Energy

the energy an object has because of its motion

light

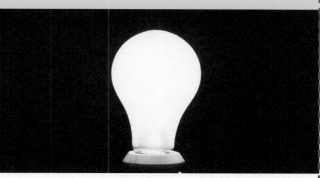

Image: Paul Fuqua

waves of electromagnetic energy; electromagnetic energy that people can see

mass

Image: Pixabay

the amount of matter in an object

motion

Image: Discovery Communications, Inc.

a change in the position of an object compared to another object

nuclear energy

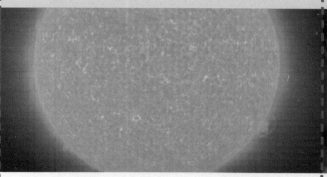

Image: Paul Fuqua

the energy released when the nucleus of an atom is split apart (during fission) or combined with another nucleus (during fusion)

photosynthesis

Image: Paul Fuqua

the process in which plants and some other organisms use the energy in sunlight to make food

potential energy

Image: wellphoto / Shutterstock.com

the amount of energy that is stored in an object; energy that an object has because of its position relative to other objects

resistance

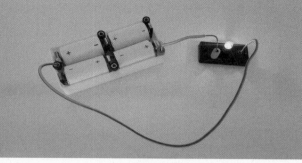

Image: fen deneylm / Shutterstock.com

the tendency of materials to not transfer energy

rotate

Image: Free-Photos / Pixabay

turning around on an axis; spinning

solar energy

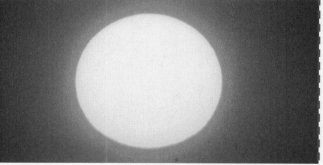

Image: Paul Fuqua

energy that comes from the sun

speed

Image: Discovery Communications, Inc.

distance traveled per unit of time

English ——— A ——— Español

acceleration to increase speed	**aceleración** aumentar la rapidez
adaptation something a plant or animal does to help it survive in its environment (related word: adapt)	**adaptación** algo que hace una planta o un animal para sobrevivir en su medio ambiente (palabra relacionada: adaptar)
air the part of the atmosphere closest to Earth; the part of the atmosphere that organisms on Earth use for respiration	**aire** parte de la atmósfera más cercana a la Tierra; la parte de la atmósfera que los organismos que habitan la Tierra utilizan para respirar
amplitude height or "strength" of a wave	**amplitud** altura o "magnitud" de una onda
analog one continuous signal that does not have any breaks	**analógico** señal continua que no tiene ninguna interrupción

antenna
a device that receives radio waves and television signals

antena
dispositivo que recibe ondas de radio y señales de televisión

Arctic
being from an icy climate, such as the north pole

ártico
que pertenece a un clima helado, como el del polo norte

— B —

behavior
all of the actions and reactions of an animal or a person (related word: behave)

conducta
todas las acciones y reacciones de un animal o una persona (palabra relacionada: comportarse)

brain
the main control center in an animal body; part of the central nervous system

cerebro
principal centro de control en el cuerpo de un animal; parte del sistema nervioso central

— C —

camouflage
the coloring or patterns on an animal's body that allow it to blend in with its environment

camuflaje
color o patrones del cuerpo de un animal que le permiten confundirse con su medio ambiente

canyon

a deep valley carved by flowing water

cañón

valle profundo labrado por el flujo del agua

chemical energy

energy that can be changed into motion and heat

energía química

energía que se puede cambiar a movimiento y calor

chemical weathering

changes to rocks and minerals on Earth's surface that are caused by chemical reactions

meteorización química

cambios en las rocas y minerales de la superficie de la Tierra causados por reacciones químicas

code

a way to communicate by sending messages using dots and dashes

código

forma de comunicarse enviando mensajes con puntos y rayas

collision

the moment where two objects hit or make contact in a forceful way

colisión

el momento en el que dos objetos chocan o hacen contacto de forma contundente

conduction

when energy moves directly from one object to another

conducción

cuando la energía pasa en forma directa de un objeto a otro

conservation of energy

energy can not be created or destroyed, it can only be changed from one form to another, such as when electrical energy is changed into heat energy

conservación de la energía

la energía no se puede crear o destruir; solo se puede cambiar de una forma a otra, como cuando la energía eléctrica cambia a energía térmica

conserve

to protect something, or prevent the wasteful overuse of a resource

conservar

proteger algo o evitar el uso excesivo e ineficiente de un recurso

convert (v)

to change forms

convertir (v)

cambiar de forma

Discovery EDUCATION

D

delta

a fan-shaped mass of mud and other sediment that forms where a river enters a large body of water

delta

masa de barro y otros sedimentos parecida a un abanico, que se forma donde un río ingresa a un gran cuerpo de agua

deposition

laying sediment back down after erosion moves it around

sedimentación

volver a depositar sedimentos una vez que la erosión los arrastra

digestive system

the body system that breaks down food into tiny pieces so that the body's cells can use it for energy

sistema digestivo

sistema del cuerpo que descompone alimentos en pequeños trozos para que las células del cuerpo puedan usarlos para obtener energía.

digital

a signal that is not continuous and is made up of tiny separate pieces

digital

una señal que no es continua y está compuesta por diminutas partes separadas

disease
a condition that disrupts processes in the body and usually causes an illness

enfermedad
afección que perturba los procesos del cuerpo

dune
a hill of sand created by the wind

duna
colina de arena creada por el viento

E

ear
organ for hearing

oído
órgano para oír

Earth
the third planet from the sun; the planet on which we live (related words: earthly; earth – meaning soil or dirt)

Tierra
tercer planeta desde el Sol; planeta en el cual vivimos (palabras relacionadas: terrenal; tierra en el sentido de suelo o suciedad)

earthquake
a sudden shaking of the ground caused by the movement of rock underground

terremoto
repentina sacudida de la tierra causada por el movimiento de rocas subterráneas

ecosystem
all the living and nonliving things in an area that interact with each other

ecosistema
todos los seres vivos y objetos sin vida de un área, que se interrelacionan entre sí

electromagnetic spectrum
the full range of frequencies of electromagnetic waves

espectro electromagnético
rango completo de frecuencias de las ondas electromagnéticas

energy
the ability to do work or cause change; the ability to move an object some distance

energía
capacidad de hacer un trabajo o producir un cambio; capacidad de mover un objeto a cierta distancia

energy source
where a form of energy begins

fuente de energía
origen de una forma de energía

energy transfer

the transfer of energy from one organism to another through a food chain or web; or the transfer of energy from one object to another, such as heat energy

transferencia de energía

transmisión de energía de un organismo a otro a través de una cadena o red alimentaria; o transmisión de energía de un objeto a otro, como por ejemplo la energía calórica

engineer

Engineers have special skills. They design things that help solve problems.

ingeniero

Los ingenieros poseen habilidades especiales. Diseñan cosas que ayudan a resolver problemas.

environment

all the living and nonliving things that surround an organism

medio ambiente

todos los seres vivos y objetos sin vida que rodean a un organismo

erosion

the removal of weathered rock material. After rocks have been broken down, the small particles are transported to other locations by wind, water, ice, and gravity

erosión

eliminación de material rocoso desgastado. Después de descomponerse las rocas, el viento, el agua, el hielo y la gravedad transportan las partículas pequeñas a otros lugares

erupt

the action of lava coming out of a hole or crack in Earth's surface; the sudden release of hot gasses or lava built up inside a volcano (related word: eruption)

erupción

acción de la lava que sale de un agujero o cráter de la superficie de la Tierra; repentina liberación de gases calientes o lava que se acumulan en el interior de un volcán (palabra relacionada: erupción)

extinct

describes a species of animals that once lived on Earth but which no longer exists (related word: extinction)

extinto

palabra que hace referencia a una especie de animales que habitaba antiguamente la Tierra, pero que ya no existe (palabra relacionada: extinción)

F

fault

a fracture, or a break, in the Earth's crust (related word: faulting)

falla

fractura, o quiebre, en la corteza de la Tierra (palabra relacionada: fallas)

feature

things that describe what something looks like

rasgo

cosas que describen cómo se ve algo

force

a pull or push that is applied to an object

fuerza

acción de atraer o empujar que se aplica a un objeto

forecast

(v) to analyze weather data and make an educated guess about weather in the future; (n) a prediction about what the weather will be like in the future based on weather data

pronosticar / pronóstico

(v) analizar los datos del tiempo y hacer una conjetura informada sobre el tiempo en el futuro; (s) predicción sobre cómo será el tiempo en el futuro con base en datos

fossil fuels

fuels that come from very old life forms that decomposed over a long period of time, like coal, oil, and natural gas

combustibles fósiles

combustibles que provienen de formas de vida muy antiguas que se descompusieron en el transcurso de un período de tiempo largo, como el carbón, el petróleo y el gas natural

friction
a force that slows down or stops motion

fricción
fuerza que desacelera o detiene el movimiento

fuel
any material that can be used for energy

combustible
todo material que puede usarse para producir energía

 G

generate
to produce by turning a form of energy into electricity

generar
producir convirtiendo una forma de energía en electricidad

geothermal
heat found deep within Earth

geotérmica
calor que se encuentra en la profundidad de la Tierra

glacier
a large sheet of ice or snow that moves slowly over Earth's surface

glaciar
gran capa de hielo o nieve que se mueve lentamente sobre la superficie de la Tierra

gravitational potential energy
energy stored in an object based on its height and mass

energía potencial gravitacional
energía almacenada debida a la ubicación en un campo gravitacional

gravity
the force that pulls an object toward the center of Earth (related word: gravitational)

gravedad
fuerza que empuja a un objeto hacia el centro de la Tierra (palabra relacionada: gravitacional)

— H —

heart
the muscular organ of an animal that pumps blood throughout the body

corazón
órgano muscular de un animal que bombea sangre a través del cuerpo

heat
the transfer of thermal energy

calor
transferencia de energía térmica

hibernate
to reduce body movement during the winter in an effort to conserve energy (related word: hibernation)

hibernar
reducir el movimiento del cuerpo durante el invierno con la finalidad de conservar la energía (palabra relacionada: hibernación)

---- I ----

information
facts or data about something; the arrangement or sequence of facts or data

información
hechos o datos sobre algo; la organización o secuencia de hechos o datos

---- K ----

kinetic energy
the energy an object has because of its motion

energía cinética
energía que posee un objeto a causa de su movimiento

landform

a large natural structure on Earth's surface, such as a mountain, a plain, or a valley

accidente geográfico

estructura natural grande que se encuentra en la superficie de la Tierra, como una montaña, una llanura o un valle

lava

molten rock that comes through holes or cracks in Earth's crust that may be a mixture of liquid and gas but will turn into solid rock once cooled

lava

roca fundida que sale por orificios o grietas en la corteza terrestre, y que puede ser una mezcla de líquido y gas pero se convierte en roca sólida al enfriarse

light

a form of energy that moves in waves and particles and can be seen

luz

forma de energía que se desplaza en ondas y partículas y que puede verse

M

magma
melted rock located beneath Earth's surface

magma
roca fundida que se encuentra debajo de la superficie de la Tierra

magnetic field
a region in space near a magnet or electric current in which magnetic forces can be detected

campo magnético
región en el espacio cerca de un imán o de una corriente eléctrica, donde pueden detectarse fuerzas magnéticas

map
a flat model of an area

mapa
modelo plano de un área

mass
the amount of matter in an object

masa
cantidad de materia que hay en un objeto

matter
material that has mass and takes up some amount of space

materia
material que tiene masa y ocupa cierta cantidad de espacio

meander
winding or indirect movement or course

meandro
movimiento o curso serpenteante o indirecto

migration

the movement of a group of organisms from one place to another, usually due to a change in seasons

migración

desplazamiento de un grupo de organismos de un lugar a otro, generalmente debido a un cambio de estaciones

---

model

a drawing, object, or idea that represents a real event, object, or process

modelo

dibujo, objeto o idea que representa un suceso, objeto, o proceso real

---

motion

when something moves from one place to another (related words: move, movement)

movimiento

cuando algo pasa de un lugar a otro (palabra relacionada: mover, desplazamiento)

---

mountain

an area of land that forms a peak at a high elevation (related term: mountain range)

montaña

área de tierra que forma un pico a una elevación alta (palabra relacionada: cadena montañosa)

nerve
a cell of the nervous system that carries signals to the body from the brain, and from the body to the brain and/or spinal cord

nervio
célula del sistema nervioso que lleva señales al cuerpo desde el cerebro, y desde el cuerpo al cerebro y/o médula espinal

nonrenewable
once it is used, it cannot be made or reused again

no renovable
una vez usado, no puede rehacerse o reutlizarse

nonrenewable resource
a natural resource of which a finite amount exists, or one that cannot be replaced with currently available technologies

recurso no renovable
recurso natural del cual existe una cantidad finita, o que no puede remplazarse con las tecnologías actualmente disponibles

nuclear energy
the energy released when the nucleus of an atom is split apart or combined with another nucleus

energía nuclear
energía liberada cuando el núcleo de un átomo se divide o combina con otro núcleo

O

ocean
a large body of salt water that covers most of Earth

océano
gran cuerpo de agua salada que cubre la mayor parte de la Tierra

opaque
describes an object that light cannot travel through

opaco
describe un objeto que la luz no puede atravesar

organ
a group of tissues that performs a complex function in a body

órgano
conjunto de tejidos que realizan una función compleja en el cuerpo

organism
any individual living thing

organismo
todo ser vivo individual

P

photosynthesis
the process in which plants and some other organisms use the energy in sunlight to make food

fotosíntesis
proceso por el cual las plantas y algunos otros organismos usan la energía del Sol para producir alimentos

pollute

to put harmful materials into the air, water, or soil (related words: pollution, pollutant)

contaminar

poner materiales perjudiciales en el aire, agua o suelo (palabras relacionadas: contaminación, contaminante)

pollution

when harmful materials have been put into the air, water, or soil (related word: pollute)

contaminación

cuando se introducen materiales perjudiciales en el aire, el agua o el suelo (palabra relacionada: contaminar)

potential energy

the amount of energy that is stored in an object; energy that an object has because of its position relative to other objects

energía potencial

cantidad de energía almacenada en un objeto; energía que tiene un objeto debido a su posición relativa con otros objetos

predator

an animal that hunts and eats another animal

depredador

animal que caza y come a otro animal

predict
to guess what will happen in the future (related word: prediction)

predecir
adivinar qué sucederá en el futuro (palabra relacionada: predicción)

prey
an animal that is hunted and eaten by another animal

presa
animal que es cazado y comido por otro

pupil
the black circle at the center of an iris that controls how much light enters the eye

pupila
círculo negro en el centro del iris que controla cuánta luz entra al ojo

R

radiant energy
energy that does not need matter to travel; light

energía radiante
energía que no necesita de la materia para viajar; luz

radiation
electromagnetic energy (related word: radiate)

radiación
energía electromagnética (palabra relacionada: irradiar)

receptor
nerves located in different parts of the body that are especially adapted to receive information from the environment

reflect
light bouncing off a surface (related word: reflection)

reflex
an automatic response

refract
to bend light as it passes through a material (related word: refraction)

remote (adj)
to be operated from a distance

receptor
nervios ubicados en diferentes partes del cuerpo que están especialmente adaptados para recibir información del medio ambiente

reflejar
rebotar la luz sobre una superficie (palabra relacionada: reflexión)

reflejo
respuesta automática

refractar
torcer la luz cuando pasa a través de un material (palabra relacionada: refracción)

remoto (adj)
que se opera a distancia

renewable
to reuse or make new again

renovable
reutilizar o volver a hacer de nuevo

renewable resource
a natural resource that can be replaced

recurso renovable
recurso natural que puede reemplazarse

reproduce
to make more of a species; to have offspring (related word: reproduction)

reproducir
engendrar más individuos de una especie; tener descendencia (palabra relacionada: reproducción)

resistance
when materials do not let energy transfer through them

resistencia
cuando los materiales no permiten la transferencia de energía a través de ellos

resource
a naturally occurring material in or on Earth's crust or atmosphere of potential use to humans

recurso
material que se origina de forma natural en o sobre la corteza o la atmósfera de la Tierra, que es de uso potencial para los seres humanos

rotate

turning around on an axis;
spinning (related word: rotation)

rotar

girar sobre un eje; dar vueltas
(palabra relacionada: rotación)

––––––––––––– S –––––––––––––

satellite

a natural or artificial object that
revolves around another object
in space

satélite

objeto natural o artificial que gira
alrededor de otro objeto en el
espacio

sediment

solid material, moved by wind
and water, that settles on the
surface of land or the bottom of a
body of water

sedimento

material sólido que el viento o
el agua transportan y que se
asienta en la superficie de la
tierra o en el fondo de un cuerpo
de agua

seismic

having to do with earthquakes or
earth vibrations

sísmico

relativo a los terremotos o a las
vibraciones de la Tierra

seismic wave

waves of energy that travel through the Earth

onda sísmica

ondas de energía que se desplazan a través del interior de la Tierra

senses

taste, touch, sight, smell, and hearing (related word: sensory)

sentidos

gusto, tacto, visión, olfato y oído (palabra relacionada: sensorial)

skin

an organ that covers and protects the bodies of many animals

piel

órgano que cubre y protege el cuerpo de muchos animales

soil

the outer layer of Earth's crust in which plants can grow; made of bits of dead plant and animal material as well as bits of rocks and minerals

suelo

capa externa de la corteza de la Tierra en donde crecen las plantas; formada por pedazos de plantas y animales muertos, así como por pedazos de rocas y de minerales

solar energy
energy that comes from the sun

energía solar
energía que proviene del Sol

sound
anything you can hear that travels by making vibrations in air, water, and solids

sonido
todo lo que se puede oír, que se desplaza produciendo vibraciones en el aire, el agua y los objetos sólidos

sound wave
a sound vibration as it is passing through a material: Most sound waves spread out in every direction from their source.

onda sonora
vibración que produce el sonido cuando atraviesa un material: la mayoría se dispersa desde la fuente en todas direcciones.

speed
the measurement of how fast an object is moving

rapidez
medida de la tasa a la que se desplaza un objeto

stimulus
things in the environment that cause us to react or have a physical response

estímulo
algo en el medio ambiente que nos hace reaccionar o tener una respuesta física

stomach

a muscular organ in the body where chemical and mechanical digestion take place

estómago

órgano muscular del cuerpo donde tiene lugar la digestión química y mecánica

---

sun

any star around which planets revolve

sol

toda estrella alrededor de la cual giran los planetas

---

survive

to continue living or existing: an organism survives until it dies; a species survives until it becomes extinct (related word: survival)

sobrevivir

continuar viviendo o existiendo: un organismo sobrevive hasta que muere; una especie sobrevive hasta que se extingue (palabra relacionada: supervivencia)

---

system

a group of related objects that work together to perform a function

sistema

grupo de objetos relacionados que funcionan juntos para realizar una función

T

tectonic plate
one of several huge pieces of Earth's crust

placa tectónica
una de las muchas partes enormes de la corteza terrestre

thermal energy
energy in the form of heat

energía térmica
energía en forma de calor

tongue
an organ in the mouth that helps in eating and speaking

lengua
órgano de la boca que ayuda a comer y hablar

topographic map
a map that shows the size and location of an area's features such as vegetation, roads, and buildings

mapa topográfico
mapa que muestra el tamaño y la ubicación de características de un área, como la vegetación, las carreteras y los edificios

trait
a characteristic or property of an organism

rasgo
característica o propiedad de un organismo

transparent

describes materials through which light can travel; materials that can be seen through

transparente

describe materiales a través de los cuales puede desplazarse la luz; materiales a través de los cuales se puede ver

tsunami

a giant ocean wave (related word: tidal wave)

tsunami

ola gigante en el océano (palabra relacionada: maremoto)

——————— V ———————

valley

a low area of land between two higher areas, often formed by water

valle

área baja de tierra entre dos áreas más altas, generalmente formada por el agua

volcano

an opening in Earth's surface through which magma and gases or only gases erupt (related word: volcanic)

volcán

abertura en la superficie de la Tierra a través de la cual surgen magma y gases, o solo gases, que hacen erupción (palabra relacionada: volcánico)

W

water
a compound made of hydrogen and oxygen; can be in either a liquid, ice, or vapor form and has no taste or smell

agua
compuesto formado por hidrógeno y oxígeno; puede estar en forma de líquido, hielo o vapor y no tiene sabor ni olor

wave
a disturbance caused by a vibration; waves travel away from the source that makes them

onda
perturbación causada por una vibración que se aleja de la fuente que la origina

wavelength
the distance between one peak and the next on a wave

longitud de onda
distancia entre un pico y otro en una onda

weathering
the physical or chemical breakdown of rocks and minerals into smaller pieces or aqueous solutions on Earth's surface

meteorización
desintegración física o química de rocas y minerales en partes más pequeñas o en soluciones acuosas en la superficie de la Tierra

work

a force applied to an object over a distance

trabajo

fuerza aplicada a un objeto a lo largo de una distancia

Discovery EDUCATION

Index

© Discovery Education | www.discoveryeducation.com

R

Record Evidence Like a Scientist 40–44,
	80–84, 124–128, 166–170
Resistance 45
Rotate 17

S

Scientific Explanation 42–44, 82–84,
	126–129, 168–170
Solve Problems Like a Scientist 4–5,
	178–182
Speed 100–101, 131, 144–145

STEM in Action 45–47, 85–87, 129–131,
	171–175

T

Think Like a Scientist 56–57, 112–113

U

Unit Project 4–5, 178–182

W

Work 36, 38